T0261037

Wireless Multimedia

...A Guide to the IEEE 802.15.3™ Standard

Dr. James P. K. Gilb

Published by
Standards Information Network
IEEE Press

Trademarks and Disclaimers

Library of Congress Cataloging-in-Publication Data

Gilb, James P. K., 1965–
   Wireless multimedia: a guide to the 802.15.3 standard / by James P. K. Gilb.

   p. cm.

   Includes bibliographical references and index.
   ISBN 978-0-7381-3668-4
   1. IEEE 802.11 (Standard) 2. Multimedia systems. 3. Wireless LANs. I. Title

   TK5105.5668.G55 2003
   004.6'8—dc21

                                                                      2003050852

The Institute of Electrical and Electronics Engineers, Inc.
3 Park Avenue, New York, NY 10016-5997, USA

Yvette Ho Sang, Manager, Standards Publishing
Jennifer Longman, Managing Editor
Andrew Ickowicz, Project Editor
Linda Sibilia, Cover Designer

Review Policy

To order IEEE Press Publications, call 1-800-678-IEEE.

Print: ISBN 978-0-7381-3668-4   SP1132

See other IEEE standards and standards-related product listings at:
http://standards.ieee.org/

Trademarks

Bluetooth is a registered trademark of Bluetooth SIG, Inc.(www.bluetooth.org/).

Certicom is a trademark of Certicom Corporation (http://www.certicom.com/).

Cisco is a registered trademark of Cisco Systems, Inc. (www.cisco.com/).

Conexant is a registered trademark of Conexant Systems, Inc. (http://www.conexant.com/).

FireWire is a trademark of Apple Computer, Inc. (http://www.apple.com/).

iDEN is a trademark of Motorola, Inc. (http://www.motorola.com/).

IEEE and 802 are registered trademarks of the Institute of Electrical and Electronics Engineers, Incorporated (www.ieee.org/).

IEEE Standards designations are trademarks of the Institute of Electrical and Electronics Engineers, Incorporated (www.ieee.org/).

Intel is a registered trademark of Intel Corporation (http://www.intel.com/).

Jini is a trademark of the Sun Microsystems, Inc. (http://www.sun.com/).

Kodak is a trademark of Eastman Kodak Company (http://www.kodak.com/).

Maxim is a registered trademark of Maxim Integrated Products (http://www.maxim-ic.com/).

Memory Stick is a registered trademark of the Sony Corporation (http://www.sony.com).

Motorola is a registered trademark of Motorola, Inc. (http://www.motorola.com/).

Nextel is a trademark of the Nextel Communications, Inc. (http://www.nextel.com).

Ntru is a trademark of Ntru Cryptosystems, Inc. (http://www.ntru.com/).

Sharp is a registered trademark of Sharp Corporation (http://www.sharpusa.com).

Sony is trademark of Sony Corporation (http://www.sony.com).

Texas Instruments is a trademark of Texas Instruments Incorporated (http://www.ti.com).

Universal Plug and Play or UPnP is a trademark of Implementers Corporation (http://www.upnp-ic.org/).

Wi-Fi is a registered trademark of the Wi-Fi Alliance (http://www.wi-fi.org/).

Dedication

I would like to dedicate this work to my wife and family, who are the most important people in my life, for their love, support, encouragement and sacrifice during the completion of the standard and this book.

James P. K. Gilb

Acknowledgement

A standard like IEEE 802.15.3 is not the work of one person, but rather depends critically on the contributions of a diverse group of individuals who dedicate their time to reviewing the drafts and solving problems. I would like to thank all of the members of the 802.15.3 Task Group, the letter ballot voters of the 802.15 Working Group and the sponsor ballot voters for the hard work that went into developing and reviewing this standard.

I would like to especially thank Jim Allen (802.15 Vice Chair and 802.15.3 Vice Chair) for encouraging me to write this book and for his diligent review of the draft.

I would like to thank my family, especially my wife, for all the support in writing this book. I would not have finished this book without their love and support.

As a member of the Task Group that developed this standard, I worked with some great people. Reviewing and analyzing the technical input from these individuals was essential in helping me to write this book. I would like to thank all of the contributors to the 802.15.3 standard, particularly: Robert Heile (802.15 Chair), John R. Barr (802.15.3 Chair), Allen Heberling (MAC Committee Chair and MAC Assistant Editor), Richard Roberts (Systems Committee Chair and Layer Management Assistant Editor), Jay Bain (MAC Assistant Editor), Jeyhan Karaoguz (PHY Assistant Editor), John Sarallo (Layer Management Assistant Editor and MAC Contributing Editor), Ari Singer (Security Assistant Editor), Dan Bailey (Security Contributing Editor), Rajugopal Gubbi (MAC Contributing Editor), Knut Odman (MAC Contributing Editor), Mark Schrader (MAC Contributing Editor), and Bill Shvodian (MAC Contributing Editor).

I would also like to thank the anonymous peer reviewers of this book for their evaluation and suggestions.

Finally, I would like to thank my editor, Jennifer Longman, and the rest of the IEEE Standards Association team for all of their hard work and dedication.

Author

James P. K. Gilb received the Bachelor of Science degree in Electrical Engineering in 1987 from the Arizona State University, graduating magna cum laude. In 1989, he received the Master of Science degree in Electrical Engineering from the same institution and was named the Outstanding Graduate of the Graduate College. He received the Ph.D. degree in Electrical Engineering in 1999, also from Arizona State University. From 1993 to 1995, he worked as an Electrical Engineer at the Hexcel Corporation's Advanced Products Division, which was subsequently bought by the

Northrop Grumman Corporation, developing advanced artificial electromagnetic materials, radar absorbing materials, and radar absorbing structures. He joined the Motorola Corporation in 1995, working initially for the Government Systems Technology Group as an RFIC designer and radio system designer. In 1999, he moved to the Semiconductor Products Sector as a Technical Staff Engineer (Member of Technical Staff) where he worked on a variety of radio systems. He developed radio architectures and specifications for new products and provided input for new process development. He joined the Mobilian Corporation in 2000, as a Senior Staff Engineer, where he developed the radio architecture and wrote the specification for the RF/analog chip that supported simultaneous operation of IEEE Std 802.11 and Bluetooth. He was also responsible for the detailed design and layout for the front-end RF circuits of the chip. He is currently the Director of Radio Engineering at Appairent Technologies where he is responsible for overseeing the implementation of the complete physical layer for IEEE Std 802.15.3. He has been the Technical Editor of the IEEE 802.15.3 Task Group since 2000 and was responsible for issuing all revisions of the draft standard. He has five patents issued and many papers published in refereed journals.

Table of Contents

Introduction

This book is intended for people who want to learn more about wireless multimedia solutions based on IEEE Std 802.15.3™. This is a new IEEE standard designed to specifically address the needs of high-speed, low-cost portable multimedia devices. Because of this focus, the standard is very different from other IEEE 802® wireless standards. This book addresses four main areas of IEEE Std 802.15.3: history, applications, implementation, and justification.

The history of the 802.15.3 standard is described in Chapter 1. This chapter also includes some of the key points of the IEEE standards process, especially as it relates to the development of 802.15.3. The 802 Task Groups use a cryptic document naming system, which is explained in the chapter as well. This will make it easier for people who are new to IEEE standards to find information from the Task Group's web site.

The applications of IEEE Std 802.15.3 are discussed in Chapter 2. While a wide variety of applications are enabled with 802.15.3, the design of the protocol focused on two areas: high-throughput and low-latency. For some applications, a steady, high-throughput connection is all that is required. Other applications require a predictable, low latency, but do not require quite as much bandwidth. While the focus of 802.15.3 is on quality of service (QoS) support, it also handles asynchronous applications efficiently.

Chapter 3 provides a high-level overview of the standard. Following the overview, the medium access control (MAC) is described in detail in Chapters 4, 5, and 6. Chapter 4 covers the operation of the MAC and includes message sequence charts that illustrate the various operations of the protocol. Chapter 5 describes the operation of dependent piconets, which can be used to improve the coexistence of 802.15.3 piconets as well as coexistence with other wireless networks. Chapter 6 completes the discussion of the MAC functionality by providing a description of the optional security methods supported by the standard.

The physical layer (PHY) is discussed in detail in Chapter 7. In addition to a discussion of the characteristics of the PHY, this chapter provides additional information regarding the anticipated performance of the PHY in typical applications. At the higher data rates supported by 802.15.3, delay spread is an important factor in the operation of the PHY. Chapter 7 provides references to justify the delay-spread numbers used in the selection criteria for 802.15.3. The chapter finishes with a survey of some radio architectures that may be useful for the implementation of highly-integrated, low-cost radio frequency integrated circuits (RFICs) for 802.15.3.

Chapter 8 discusses the interfaces that were defined in IEEE Std 802.15.3. These interfaces support connections with higher layers in the protocol stack, e.g., IEEE Std 802.2™ and IEEE Std 1394™. Although the interfaces are optional to implement, they do provide insight into the operation of the protocol.

The final chapter in the book, Chapter 9, deals with the issue of coexistence. Because unlicensed allocations do not guarantee clear radio spectrum, wireless devices need to be able to handle the presence of interferers from a wide variety of sources. In the 2.4 GHz band, these include microwave ovens, IEEE Std 802.11b™ and IEEE Std 802.11g™ wireless local area networks (WLANs), IEEE Std 802.15.1™ personal area networks (PANs), etc. Simulation results for 802.15.3 coexisting with these types of interferers are presented in the chapter. In addition, the techniques provided by 802.15.3 to enhance the coexistence with these networks is also discussed in the chapter.

Throughout the book, additional information is given that provides justification for the choices made in the standard. The development process was iterative and so some ideas were introduced, only to be discarded later as further analysis showed problems with the proposed solutions. In some cases, functionality was cut from the standard, not because it was broken, but rather that its marginal utility was too low to justify the added complexity.

Acronyms and Abbreviations

The following acronyms and abbreviations are used in this book:

ADC	analog-to-digital converter
AES	advanced encryption standard
APS	asynchronous power save
ATP	association timeout period
BcstID	broadcast identifier
BER	bit error ratio
BPSK	biphase-shift keying
BSID	beacon source identifier
CAP	contention access period
CCA	clear channel assessment
CFA	call for applications
CSMA/CA	carrier sense multiple access/collision avoidance
CTA	channel time allocation
CTAP	channel time allocation period
CTRq	channel time request
CTRqB	channel time request block
CWB	continued wake beacon
DAC	digital-to-analog converter.
DBPSK	differential biphase shift keying
DES	data encryption standard
DEV	device
DEVID	device identifier
DestID	destination identifier
DSSS	direct-sequence spread spectrum
DSPS	device synchronized power save
DQPSK	differential quadrature phase-shift keying
D8PSK	differential eight phase-shift keying
EVM	error vector magnitude

FCS	frame check sequence
FER	frame error ratio
FSK	frequency shift keying
HCS	header check sequence
IE	information element
IF	intermediate frequency
ISI	intersymbol interference
LO	local oscillator
LQI	link quality indication
MAC	medium access control
MCTA	management channel time allocation
MSC	message sequence chart
MSDU	MAC service data unit
OFDM	orthogonal frequency division multiplexing
OID	object identifier
PAN	personal area network
PCTM	pending channel time map
PHY	physical layer
PIB	PAN information base
PNC	piconet controller
PNCID	piconet controller identifier
PNID	piconet identifier
ppm	parts per million
PS	power save
PSD	power spectral density
PSK	phase-shift keying
PSPS	piconet synchronized power save
PVR	personal video recorder
QAM	quadrature amplitude modulation
QPSK	quadrature phase-shift keying
RFIC	radio frequency integrated circuit
RSSI	receiver signal strength indication
SDL	specification description language

SNR	signal-to-noise ratio
SPS	synchronized power save
SrcID	source identifier
TCM	trellis coded modulation
TDMA	time division multiple access
UnassocID	unassociated ID
UWB	ultra-wideband
VGA	variable gain amplifier
VLIF	very low intermediate frequency
ZIF	zero intermediate frequency

Chapter 1 Background and History

This book provides an introduction and explanation of the IEEE 802.15.3™ wireless personal area networks (WPANs) standard. The goal of the book is to address three areas:

a) Clarify IEEE Std 802.15.3 for individuals who are implementing compliant devices;

b) Show how IEEE Std 802.15.3 can be used to develop wireless multimedia applications;

c) Provide insight into the reasons for the choices made in developing IEEE Std 802.15.3.

This chapter provides an introduction to IEEE standards, the IEEE 802.15™ family of standards, and the reasons why IEEE Std 802.15.3 was developed.

WHAT IS AN IEEE STANDARD?

IEEE Standard 802.15.3 is part of the family of standards developed by the Institute of Electrical and Electronic Engineers, Inc. (IEEE). In particular, it is a part of the Local Area Network/Metropolitan Area Network (LAN/MAN) standards activity of the IEEE Computer Society. These

 The IEEE develops a variety of standards in many different areas other than the LAN/MAN standards, which are sponsored by the IEEE's Computer Society.

Visit http://standards.ieee.org and http://grouper.ieee.org/groups/index.html for more information.

standards allow different manufacturers to create interoperable networking equipment, such as Ethernet (IEEE Std 802.3™), Token Ring (IEEE Std 802.5™), and Wi-Fi® (IEEE Std 802.11b™). Interoperability is a key requirement for networking equipment because its value comes from its ability to communicate with other equipment. Network equipment, by definition, is not useful by itself. At a minimum, two compatible devices are required to enable communications.

A standard helps the user by providing choices for purchasing hardware and keeps prices down by encouraging competition. Networking equipment that conforms to a standard also helps to ensure that the consumer will not be left with an orphaned solution. With proprietary solutions, a decision by a single manufacturer can kill a product and force consumers to change all of their equipment at a potentially large cost or incur costly service fees to maintain obsolete equipment. On the other hand, if there is a market for the equipment compliant to a standard, then it is possible for any manufacturer to provide the solution.

Manufacturers also benefit from the technical efforts that go into developing standards. The development of an IEEE standard involves thousands of hours of top-quality engineering time from many different sources in the proposal, development, and review of the standard. Although the competition inherent in the production of equipment compliant with a standard will keep prices low, this can actually help the manufacturer. While the per-unit pricing will generally be lower for a standardized product, the overall market will be correspondingly larger, providing greater profit potential for the manufacturer. Because a standard prevents proprietary lock-in, consumers tend to adopt these products more rapidly and so the total market often grows faster for a standards-compliant solution than for a proprietary solution.

The product design team benefits as well by developing a product that implements a standard. The standard development process includes an extensive peer review that tends to find almost all of the errors in the standard, so that the design engineer can focus on implementation rather than debugging the design of the protocol. A standard also allows suppliers to develop solutions that will meet the design engineer's needs.

THE 802.15 FAMILY

The IEEE 802 standards activities are divided into many different Working Groups (802.3, 802.11, 802.15, 802.16, etc.) that deal with a single type of networking standard. The 802.15 Working Group is tasked with developing standards and recommended practices for personal area networks (PANs). As of November 2003, within the 802.15 Working Group, there were five Task Groups that were either currently working on standards or had finished their work.

The first 802.15 project, IEEE Std 802.15.1™, was designed to provide a vehicle for creating a standard out of the Bluetooth® specification. The IEEE standards development process is an open one that allows input from any interested party. The Bluetooth

 The SDL model in IEEE Std 802.15.1 does not correspond to the text in the standard or the Bluetooth 1.1 specification. The model was created for an older version of the specification to prove to the SIG that there were significant problems with the Bluetooth 1.0x specifications.

specification, however, was developed by a small group of companies called the Bluetooth Special Interest Group (SIG), who retained control of the final specification. While much of the 802.15.1 standard is a word-for-word version of the Bluetooth specification, the 802.15.1 Task Group did provide additional material for the standard, including a specification definition language (SDL) model, an interface to the 802.2™ logical link control (LLC), and other text to help explain the specification. IEEE Std 802.15.1 was approved in April 2002 by the IEEE Standards Association (IEEE-SA), and the Task Group is no longer active.

Soon after the work on IEEE Std 802.15.1 began, the 802.15 and 802.11 Working Groups became concerned about the ability of 802.11b and 802.15.1 networks to coexist when they were operating in close proximity. The 802.15 Working Group then formed a new Task Group, 802.15.2, to develop a Recommended Practice that would improve the ability of the two wireless networks to coexist. A Recommended Practice is like a standard, except that it contains suggestions rather than requirements for interoperability. The suggestions represent "best practices" that can be used by implementers to improve their products. The IEEE 802.15.2 Recommended Practice was approved by both the 802.11 and 802.15 Working Groups and in June of 2003, the IEEE Std 802.15.2 was approved IEEE-SA.

The 802.15.3 Task Group was formed to create a standard that would enable wireless connectivity of high-speed, multimedia-capable portable consumer devices. The goals were to get greater than 20 Mb/s data rate while maintaining quality of service (QoS) for the data streams, low-power, and low-cost. IEEE Std 802.15.3, which is the subject of this book, was approved by the IEEE-SA in June 2003.

The 802.15.4 Task Group also took on the goal of very low-cost and low-power wireless networks. This group was tasked with developing a standard for low data rate devices (<250 kb/s) at a cost that was less than either IEEE Std 802.15.3 or IEEE Std 802.15.1. The 802.15.4 standard was approved by the IEEE-SA in May 2003.

The final group within 802.15 is the 802.15.3a™ Task Group. Their goal is to develop a new physcial layer (PHY) that will go with the 802.15.3 medium access control (MAC) to enable data

 New Study Groups and Task Groups are formed in the 802.15 Working Group from time to time. Refer to the 802.15 web page (http://grouper.ieee.org/802/15/index.html) for the latest information.

rates of greater than 110 Mb/s to more than 400 Mb/s. The Task Group has completed the call for applications (CFA) and the call for proposals (CFP) and has begun the downselection process. Once a single proposal has been selected and confirmed by a 75% vote of the Task Group and Working Group, the process of writing the draft standard will begin.

WHY 802.15.3?

IEEE Std 802.15.3 was specifically written to address a class of applications that did not have a wireless standard. Some of the wireless applications that were submitted in response to the call for applications for IEEE Std 802.15.3 were as follows:

- Connecting digital still cameras to printers or kiosks;

- Laptop to projector connection;

- Connecting a personal digital assistant (PDA) to a camera or a PDA to a printer;

- Speakers in a 5.1 surround-sound system connecting to the receiver;

- Video distribution from a set-top box or cable modem;

- Sending music from a CD or MP3 player to headphones or speakers;

- Video camera display on a television;

- Remote viewfinders for video or digital still cameras.

These applications are mainly in the consumer electronics arena and have the following characteristics:

*   *High throughput requirement:* greater than 20 Mb/s to support video and/or multi-channel audio;

*   *Low power usage:* to be useful in battery-powered portable devices;

*   *Low cost:* to add value to consumer electronic devices;

*   *Guaranteed bandwidth:* to provide QoS for applications sensitive to latency;

*   *Ad hoc connectivity:* to provide the benefits of networking without unduly restricting portability or mobility;

*   *Simple connectivity:* to make networking easy and eliminate the need for a technically sophisticated user;

*   *Independent connectivity:* because there is value in networking devices together even without a connection to the Internet;

*   *An assurance of privacy:* to assure the user that only the intended recipients can understand what is being transmitted.

Because the existing wireless standard, 802.11™, was created to address the delivery of the bursty, asynchronous traffic that is common in a LAN, it did not address the special requirements of these applications. Although 802.11 provides 54 Mb/s of instantaneous data rate with either 802.11a™ and 802.11g™, under ideal conditions the actual throughput is less than 31 Mb/s for 802.11a and is less than 30 Mb/s for 802.11g [B32]. The throughput of 802.11g drops to less than 18 Mb/s if there are any 802.11b stations in the wireless local area network (WLAN) [B32]. In addition, the 802.11 MAC does not allow for the consistent reservation of channel time that is a requirement for applications that need QoS. Networks that use 802.11 are mainly set up by professionals in businesses, hotels, airports, coffee shops, and so on, where a networking professional is able to handle the complexities of the network. Although IEEE Std 802.11 supports ad-hoc networking, this mode is difficult to set up and it does not support power-save modes for the devices in the network. Users generally rely on the infrastructure mode, which does not allow direct peer-to-peer connections or ad-hoc networks and requires a dedicated access point (AP).

This is not to say that IEEE Std 802.11 is not a good protocol. It is an excellent solution for data traffic that is typical on WLANs. It allows stations to quickly get access to the medium to send data without going through a channel time allocation process. It has routing and roaming capabilities that enable users to be mobile while maintaining a connection to the infrastructure. The 802.11 security model for centralized authentication is consistent with its use in extending the wired infrastructure to wireless stations. Network adapters for 802.11b have fallen dramatically in price as well and they are now cheaper than Bluetooth network adapters while providing 11 times the raw data rate[1].

Although IEEE Std 802.11 is able to solve some of the consumer electronics applications under certain configurations and traffic models, it does not solve the broad range of connectivity and QoS requirements for a growing class of existing and emerging portable, multimedia consumer electronic devices. Creating a solution for these applications required more than just a new PHY, it also required a completely new MAC to create a protocol that would be optimized for the consumer electronics space.

HISTORY OF 802.15.3

Multiple groups have worked on the concept of high-speed, wireless connectivity for portable, multimedia devices. The HomeRF[2] Working Group began work on a high-speed multimedia-capable specification in 1999, called HomeRF Multi-Media (or HRF-MM), that was aimed at >20 Mb/s data rates. The HomeRF multimedia specification was never released, however, the group did use a new Federal Communications Commission (FCC) report and order that allowed wideband frequency hopping in the 2.4 GHz band to extend the data rates of HomeRF 1.1 from 2 Mb/s to 10 Mb/s.

In early 2000, the Bluetooth SIG also began work on a higher speed extension for Bluetooth 1.0, which was called Radio 2. The initial goal was to achieve a data rate of least 10 Mb/s while maintaining backward compatibility with Bluetooth 1.x implementations. Because of the closed development process used by the Bluetooth SIG, little information is publicly available about the status of this

---

[1]   Products that use 802.11g are also now available, which offer 28 times the throughput of Bluetooth and are as cheap or cheaper than Bluetooth network adapters.

[2]   The HomeRF Working Group disbanded in January of 2003.

effort. By the middle of 2003, the Bluetooth SIG had announced that a smaller increase in the data rate, from 1 Mb/s to 2 or 3 Mb/s, would be available first with a 12 Mb/s enhancement that would be available later. It is not clear, however, how Bluetooth Radio 2 devices will be able to compete with existing low-cost WLANs[3]. In addition, it has been reported that the high-rate Bluetooth Specification will require an entirely different MAC and PHY from what is currently used in Bluetooth 1.x implementations [B31]. If this difference remains in the final revision of the Radio 2 specification, then it is unlikely that a combination Bluetooth 1.x and Radio 2 implementation would be less expensive than a combination Bluetooth1.x and 802.15.3 solution or even a combination Bluetooth 1.x and 802.11g solution.

In late 1999, Kodak™, Motorola®, and Cisco® approached the IEEE Std 802.15 Working Group to request a new study group to look at wireless MAC and PHY that would support portable, multimedia-capable devices. Unlike the HomeRF and Bluetooth efforts, the IEEE uses an open process that allows anyone to present a proposal for the development of a standard. The Study Group became a Task Group at the March 2000 meeting in Albuquerque, NM, USA. At that meeting, there were already responses to the call for applications. These applications were used to develop a criteria document that set out the requirements for the new standard in three main areas: system, MAC, and PHY. Once the criteria document was finished, there was a call for proposals that resulted in four MAC and nine PHY proposals. The topics of each of the four MAC proposals are listed in Table 1–1.

Table 1–1:   IEEE Std 802.15.3 MAC proposals

Description	Document number
Modified HiperLAN/2	00/196r6
802.11 "lite" with time slot additions	02/205r2 and 00/218r1
Bluetooth-like with extensions	00/212r5 and 00/213r0
802.11 "lite"	00/197r2

---

[3] EE Times reported that "Still, skeptics wonder why anybody needs a high-rate Bluetooth when consumers can safely turn to Wi-Fi for faster wireless file transfer or data streaming." [B31]

A note about the document numbering used for 802.15.3 is in order. The 802.15 Working Group has had two different naming schemes for documents. The changeover in the document naming scheme occured during the September 2003 meeting. Documents prior to 2003 and those in 2003 with sequence numbers less than 300 use the older numbering scheme. Documents that have been submitted to the 802.15 Working Group are all publicly available on the 802.15 website, http://grouper.ieee.org/groups/802/15/index.html, in the "Doc archives" link.

A typical document name in the older document naming method is "00196r6P802-15_TG3-HiPeRPAN-merged-proposal.pdf." The first two digits are for the year followed by a three-digit number assigned sequentially by the Chair of the Working Group followed by the letter "r" and the revision number. All of the 802.15.3 documents contain "P802-15_TG3-" followed by the title. The "TG3" indicates Task Group 3, which is 802.15.3. Thus, the example document name indicates that it is the sixth

> The document numbering convention for each of the 802 Working Groups is different. The best place to find this information is in the Operating Rules of the Working Group. The Working Group rules, as well as the document archive for the Working Group can be found on their web pages. Links to all 802 working groups can be found on http://ieee802.org/dots.html

revision of 196th document number assigned by the Chair of 802.15 in the year 2000 for Task Group 3 entitled "HiPeRPAN merged proposal" in pdf format. The parts of an 802.15 document name are illustrated in Figure 1–1.

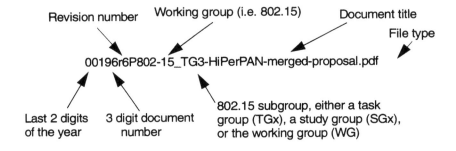

Figure 1–1: Old (prior to September 2003) IEEE 802.15 document naming conventions

A typical document using the new document naming method is "15-03-0516-01-0030-Draft-editorial-and-enhancements.doc." The string of 14 digits is called the Document Control Number (DCN). The first two digits indicate the Working Group (WG 15) while the next two digits indicate the year that the document was submitted, i.e., 2003. The next four digits are a number assigned sequentially by the document management software to uniquely identify the document. This is followed by a two digit revision number and a four character Task Group identifier followed by the title.[4] Thus the example document name indicates that it is the first revision of the 516th document submitted in 2003 for Task Group 3 entitled, "Draft editorial and enhancements" in doc format. The parts of a new 802.15 document name are illustrated in Figure 1–2.

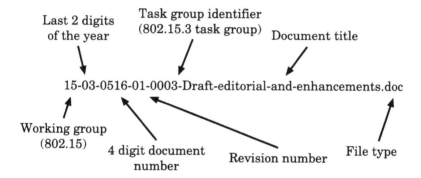

Figure 1–2:   Current IEEE 802.15 document naming conventions

In the down-selection process, one of the proposals (00/212r5, 00/213r0) was withdrawn and the other three agreed to a compromise outline of the protocol that used a beacon followed by a contention access period (CAP) followed by assigned time slots. This merged proposal received 100% approval from the Task Group. The PHY selection, however, was much more contentious. The nine PHY proposals are summarized in Table 1–2. The acronyms used in the table are defined on page xvii.

---

4   The Task Group identifier can be both digits and characters, e.g., "003a" is Task Group 3a while "mmwi" is the Millimeter Wave interest group.

Table 1–2: IEEE Std 802.15.3 PHY Proposals

Frequency	Modulation	Coding	Document numbers
5 GHz	4 level FSK	None	00/206r1
5 GHz	OFDM with BPSK, QPSK, 16-QAM, DQPSK and D8PSK	Turbo	00/196r6
2.5 GHz	2 level FSK, QPSK,	None	00/214r7, 00/215r0
2.5 GHz	M-ary Bi-Code Keying with QPSK and 16-QAM	Reed-Solomon	00/210r10
2.5 GHz	QPSK, 16-QAM	Turbo	00/199r2, 00/200r9
UWB	Bi-phase with DSSS	None	00/195r8
2.5 GHz	8PSK, QPSK, 16, 32, 64-QAM	TCM	00/211r5
2.5 GHz	DBPSK, DQPSK, 8-QAM, 16-QAM	Reed-Solomon	00/197r2, 00/198r3

The down-selection vote was held at the IEEE 802 plenary meeting in Tampa, FL, USA, in November 2000. At that point, two of the proposals were withdrawn (00/206r1 and 00/195r8) and the final proposal selected was a compromise of two proposals, (00/214r7, 00/215r0, 00/199r2, and 00/200r9). This final version specified the modulation but left the coding to be determined at the January 2001 meeting in Monterrey, CA, USA. At that meeting, trellis coded modulation (TCM) was selected as the coding with the base rate left as uncoded to allow simple, low-power, low-cost devices. This completed the PHY selection process and enabled the group to continue with the task of drafting the standard.

The initial drafting process proceeded informally with different authors given clauses to write and the intermediate results were posted for comment. In the November 2001 meeting in Austin, TX, USA, the group closed the last outstanding issue on the draft and was able to begin the draft approval process.

The draft approval process involves two stages of balloting: Working Group letter ballot and Sponsor letter ballot. The first balloting process seeks the approval of the entire Working Group: the 802.15 Working Group in this case. If the Working Group approves of the draft, it is forwarded to Sponsor ballot (conducted by the IEEE-SA), where interested individuals review and vote on the contents of the draft standard. The IEEE 802.15.3 draft standard began the Working Group letter ballot process in early December 2001 with a 40-day letter ballot. In the letter ballot, the voters are asked to vote yes, no, or abstain and to provide comments on things that they wanted changed in the draft. In the case of a no vote, the voter is required to provide comments that indicate the changes that would be required in order to change the person's no vote to a yes vote. The minimum approval for the ballot is 75%, but traditionally, the 802 Working Groups seek to get greater than 90% approval.

The first letter ballot failed to get 75% approval[5], but with changes to add security suites, coexistence analysis, and improved power management text, a sufficient number of no voters changed their votes to yes to enable a recirculation ballot. The draft went through three recirculation ballots, each time achieving greater than 75% approval before being approved to be sent to Sponsor ballot following the November 2002 802 plenary meeting in Kauai, HI, USA. A summary of the voting and comments for the Working Group letter ballots is given in Table 1–3.

The draft began the Sponsor ballot process in December of 2002. The first Sponsor ballot received 76% approval[6], just making the 75% requirement for passing. In response to the comments received, the Task Group made some changes to the draft that satisfied the majority of the no voters. The mandatory status of the device synchronized power save (DSPS) power-save mode had been a contentious issue since the first Working Group letter ballot. Based on a compromise that was agreed to in Kauai, the group was finally able to close out this issue.

 The IEEE SA has two numerical requirements for a ballot to be valid. First, greater than 75% of those eligible to vote must return ballots. Second, the number of abstain votes must be less than 30% of the total votes received. If the ballot is valid, it passes if 75% of the people vote yes of the total number of people voting yes or no (abstains do not count in this tally).

In addition, the Task Group was informed that the authentication protocols that were in the draft were likely outside of the scope of a MAC and PHY standard. This could have caused problems with the approval of the standard with either the

---

5 75 of the 89 eligible voters in the 802.15.3 Working Group voted in the first letter ballot.

6 There were 101 people in the balloting pool and in the first ballot, 79 people returned ballots in time (> 75% return ratio). There were 56 affirmative votes, 17 negative votes, and 6 abstentions (abstains < 30%). There were 447 technical comments and 379 editorial comments.

802 Executive Committee (802 ExCom) or the IEEE Standards Review Committee (RevCom). In order to avoid any problems and to ensure that the standard could be quickly approved, the Task Group removed the three authentication methods from the draft and attempted to purge the document of any references to authentication.

The revised (and much shorter) draft was recirculated to the Sponsor balloting group. This ballot came in with better than 96% approval[7] and the comments, much fewer this time, dealt mostly with finishing the removal of the security suites. After implementing changes to address these issues, the draft was recirculated a final time. In this final round of Sponsor balloting, the draft received 97% approval[8] and no new comments were received. The Working Group then forwarded the draft to the 802 ExCom where it was approved and then sent on to the IEEE RevCom for consideration. RevCom approved the draft in their June 2003 meeting and the IEEE Standards Board affirmed their decision the next day, June 12, 2003, making 802.15.3 an official standard.

Table 1–3: 802.15.3 Working Group letter ballot results

Ballot #	Vote results			Comments		
	Yes	No	Abstain	Technical Required	Technical	Editorial
LB12	49 (69%)	22 (31%)	4	554	612	685
LB17	64 (88%)	9 (12%)	2	444	131	622
LB19	66 (89%)	8 (11%)	1	326	72	153
LB22	70 (95%)	4 (5%)	1	62	24	159

---

[7] The first sponsor ballot recirculation result was 73 approve, 3 disapprove, and 6 abstain with 57 technical comments and 117 editorial comments.

[8] The final sponsor ballot recirculation result was 76 approve, 2 disapprove, and 6 abstain with no comments received.

Chapter 2 802.15.3 applications

IEEE Std 802.15.3, like all IEEE standards, addresses a set of applications. In the process of developing the standard, a call for applications (CFA) is sent out to solicit responses from manufacturers and users to indicate the types of applications for which they need a solution. Although not every application can be addressed with a single standard, the task group tried to enable as many applications as possible identified in the CFA while aiming for low cost, complexity, and power.

THE HIGH-RATE WPAN THEME

There are several wireless standards that can be applied to some of the following applications, however, only 802.15.3 is able to address all of the requirements for portable, wireless multimedia devices. In addition, it is important to understand the underlying philosophy of 802.15.3 in order to understand what the excitement is about. The general applications of local wireless solutions for consumer electronic devices can be divided into data networks, such as those used for enterprise WLANs or desktop internet applications, and multimedia networks, like those used for entertainment and portable devices. These networks have different purposes and requirements. After the CFA, it became clear that multimedia networks need to include a large class of devices in order to be a complete system. For example, a television set is not just a wireless video distribution opportunity. It had its own class of peripherals such as remote controls, 6.1 speakers, keyboards for on-screen Internet access, input from personal video cameras, and still cameras. In order to be useful in the home or on vacation, the needs of all these devices had to be addressed in a single standardized solution. There had to be a full systems solution, or the customer would not be satisfied.

IEEE Std 802.15.3 was a "ground up" opportunity to develop a new standard that served a large, untapped market, not accessible to Bluetooth, 802.11, and other approaches, and to serve as a platform for future enhancements.

STILL IMAGE APPLICATIONS

One of the original applications that drove the development of IEEE Std 802.15.3 was still image transfer from a digital camera to a photo kiosk. The increasing size of images in digital cameras combined with the impatience of users created a need for low-cost, high data-rate, ad-hoc wireless connection. Based on a 15-second transaction time limit for the user's attention span, the data rate needed to be at least 20 Mb/s. The wireless photo kiosk is intended to allow owners of digital cameras to create prints, photo CDs, store the picture at a central site, or even email the picture to selected individuals with a simple button push. Digital cameras are now getting the capability of adding information to the picture, when it is taken, that indicates what the user wants to do with the picture, e.g., print to paper, size of the prints, number of copies, where to store the data file, and so on.

One of the advantages of wireless image transfer is that it allows people to "unload" their digital cameras into permanent storage, prints, photo CDs, or a central Internet location, e.g., www.ofoto.com. This is especially important for people when they are are traveling on vacation. The typical digital camera user saves their pictures on their home computer so that they can erase the files from the camera or memory card. However, most people do not have laptops and even fewer take their laptops or computers on vacation, which is where most of the pictures will be taken. Thus, people are not able to use their digital cameras as much as they would use a standard film camera.

The photo kiosk application offers value to everyone involved[9]. The store or facility that houses the kiosks gets rental money from the kiosk owner as well as additional business and traffic from people that use the machine. The kiosk owner is able to collect money for printing pictures or making CDs. If the pictures are sent to an online site for storage, the kiosk owner may collect a transmittal fee and the website owner may also get money for storing the picture. The person using the camera gets to empty their flash card and receive a copy of the pictures (electronic, paper, or both). Because the camera also acts as a display device, it can be used for uploading content, e.g., music, short videos, ads, or even other pictures.

---

[9] Currently there are approximately 50,000 photo kiosks and they are expanding rapidly.

The still image transfer capabilities are also a compelling application for the home. Less than 50% of U.S. households have a computer, and yet virtually every household has a film camera[10]. In a typical house with a computer, it is located in a back room in an out of the way place and the computer is not normally turned on. When the user wants to transfer images or print them, they must go to the computer, turn it on, wait for it to boot, and then they can begin to transfer images. In order to expand the market for digital cameras and related services, there needs to be a way to use the camera without requiring the intervention of the PC.

The simple way to do this is to wirelessly enable both the printer and the camera. However, the images that the camera prints can be large and the user is not going to want to wait for a slow data link to transfer the pictures. With an IEEE Std 802.15.3 wireless link, the user can easily send the pictures to the printer. The market has already provided ink jet printers that have slots for the various types of digital camera storage, e.g., compact flash (CF), secure digital (SD) cards, multimedia cards (MMC), and Memory Stick®. As an extra cost option, they also can have a small display for viewing the pictures on the card. However, with a wireless link, the camera can be used as a display and control device for the printer. In addition, manufacturers can offer a less expensive printer because they do not have to add the expense of supporting multiple slots and formats.

Another application of the still image transfer would be the remote display of images on various devices. For example, the user could create a slide show of images and display them on an 802.15.3-enabled television. Because televisions and even high-definition televisions (HDTVs) are relatively low resolution compared with digital pictures, even a 1 megapixel camera would create an image that would display well on an HDTV. This would require application level support on the television for JPEG pictures. Users could also upload pictures to digital picture frames without having to take the frame off the wall to plug in the cable.

In summary, some of the applications that would use still image or bulk file transfer are:

- Printing pictures to a photo kiosk

- Uploading content from a kiosk (movies, ads, still photos, music, etc.)

---

[10] In 2003 there were approximately 800 million film cameras and 37 million digital cameras.

- Remote display of still images on televisions or picture frames
- Direct camera to ink jet printing in the home
- Serve as display and control device for the printer

TELEPHONE QUALITY AUDIO APPLICATIONS

In this context, audio applications are those that transmit voice between two parties in a manner similar to the telephone system. Although the data rates for this type of communications are not particularly high, about 9.6 to 64 kb/s depending on the quality required, these applications have the most stringent latency requirements. The reason for this is that although people can deal with delayed video, as long as the associated audio remains in sync, they are not used to long pauses in conversation. In most cultures, a long (>0.5 s) latency in receiving audio traffic from the remote person will cause the other individual to begin talking, often on top of the remote person's response.

The quality required for audio transmission, the latency, dynamic range, bit rates required, and so on, have been extensively studied by the telephone and cellular companies. Because of this research and the experience that people have with telephones today, the requirements for audio applications are well-documented. In many places, people are used to very high-quality voice connections, which sets a high expectation for any protocol/radio combination that seeks to replace a wired solution. Part of the design of IEEE Std 802.15.3 was aimed at solving the audio problem. The 802.2 priority mapping (see Annex A in IEEE Std 802.15.3) specifies less than 10 ms of delay and jitter for voice applications. This number is relatively low because it is expected that there will be other delays and jitter from the wireline portion of the connection. If the connection is only between wireless devices (DEVs) in the piconet, e.g., an intercom system, then the latency could be expanded to greater than 80 ms.

The latency for voice and any retransmissions, if needed, is controlled by the interval requested in the channel time request. For example, if the superframe duration is 10 ms, then the DEV could request allocations every fourth superframe and suffer no more than 40 ms latency. The maximum allowed superframe duration is 65 ms, which puts an upper limit on the latency for a single allocation in a superframe. If the DEV wanted to have a lower latency, it could request that its channel time be split into two allocations spread out in the same superframe, i.e., a super-rate allocation.

In summary, some of the applications that would use telephone quality audio are:

* Cordless telephones
* Intercoms

HIGH QUALITY AUDIO APPLICATIONS

The high-quality audio market is distinct from telephone-quality audio. In the case of telephony, the end-to-end latency is very important. A typical high-quality audio application, on the other hand, is relatively insensitive to latency over the air as long as there is sufficient buffering and bandwidth to maintain the data flow. For example, many modern mobile CD players read ahead on the CD and buffer the information to prevent skips in the playback when movement of the player causes the read process to skip. Although some buffering is allowed in this application, users still expect the sound to start soon after the "start" button is pressed and for it to continue uninterrupted for as long as they want to listen to it. In addition, the end points of an audio system, i.e., the speakers and microphones, are relatively low-cost items and so the buffer space will be constrained.

High quality audio also requires a much greater data rate than simple telephone quality audio. While wireline quality voice connections run at 64 kb/s[11], CD quality audio requires (2*44.1 kHz* 8 bits/sample) = 705.6 kb/s if it is sent uncompressed. A 5.1 surround sound system (e.g., 6 channels) would require three times the data rate, a little more than 2.1 Mb/s, to send the data uncompressed. While lossy compression methods, e.g., MP3 or Ogg Vorbis, are acceptable for portable digital audio applications, home and professional sound systems require uncompressed data to maintain greater than 90 dB dynamic range.

One of the challenges in wireless distribution of sound, especially in a 5.1 surround sound system, is to keep the playback synchronized. While the 802.15.3 standard does not specify a method for synchronizing different applications, the beacon provides a time frame of reference for all of the DEVs in the piconet.

---

[11] Digital cellular telephones use compression for voice transmission and generally use less than 19 kb/s, some as low as 9.6 kb/s.

This allows all of the DEVs to use a common timing frame of reference for playback. In addition, the standard provides sufficient flexibility for an application to create a method for synchronizing playback among a variety of DEVs.

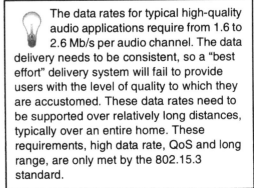 The data rates for typical high-quality audio applications require from 1.6 to 2.6 Mb/s per audio channel. The data delivery needs to be consistent, so a "best effort" delivery system will fail to provide users with the level of quality to which they are accustomed. These data rates need to be supported over relatively long distances, typically over an entire home. These requirements, high data rate, QoS and long range, are only met by the 802.15.3 standard.

In summary, some of the high-quality audio applications that are supported by 802.15.3 include:

- Digital audio players (e.g., Ogg Vorbis, MP3, etc.)

- Headphones

- Speakers, especially 5.1 surround sound systems

- Wireless microphones

GAMING APPLICATIONS

A low-cost, high-speed wireless network with guaranteed time allocations opens up new types of applications for gaming. In terms of the number of devices, the home console and portable video game market is larger than the home computer market. The most obvious application of wireless connectivity in video games is to remove the wires that connect the controllers to the game console. Although the amount of information that is transferred to and from the controller is small, the users are very sensitive to latency. A controller that fails to get its command through in a timely manner will cause the user to lose the game and will be quickly returned to the store. IEEE Std 802.15.3 allows the application to control the latency by selecting the allocation interval for the data. The data transfers are bidirectional as well because game controllers often implement physical feedback mechanisms to enhance the gaming experience.

The controllers for the game console are only a small part of the total gaming market. Once the game console has become wirelessly enabled with 802.15.3, other applications are available. For example, with multiple game controllers in a relatively small area, the users could engage in a multiplayer game. Because multiple consoles are used, the players could each have a different viewpoint of the action in their display. Game consoles now have the ability to access the Internet. In this case, the console could act as the wireless router for the house, distributing information to wireless portable devices such as web pads. The game console could also be used to play music that is sent wirelessly to speakers or headphones. At some point, the game console may take on the function of the home computer for many households.

Another application would be to enable handheld games to interact with each other wirelessly. Multiplayer games that could be formed whenever a group of people with handheld gaming devices was in range. The gaming devices are taking on new roles as well, such as providing playback of digital music. Having a wireless connection in the handheld game would allow the user to upload music, videos or images from kiosks, computers, or the game console. Because the handheld game has a screen as well as buttons, it can serve as the interface for this transaction. Not only that, but new applications will be found for the device using the display and controls to provide input to and feedback from another 802.15.3 device that, for cost reasons, does not have a user interface.

In summary, some of the possible gaming applications are:

- Wireless controllers

- Multiplayer gaming with home console

- Game console as Internet portal

- Handheld multiplayer gaming

- Digital music, video, or image upload to handheld games

VIDEO AND MULTIMEDIA APPLICATIONS

Multimedia applications have the highest requirements for throughput, but they often have more relaxed latency requirements. Streaming multimedia delivery over the Internet is accomplished today with buffering. This model works well when the endpoint is a computer that has a large hard drive and RAM. However, many consumer devices do not have large amounts of RAM or non-volatile storage available. Thus, while the latency requirement is relaxed for streaming multimedia applications, there is still an upper bound on the latency that is permissible for these applications.

The most commonly cited streaming multimedia application is the set-top box application. Although satellite and cable television are normally delivered to a single box in a house, the typical home has more than one television. The carriers (e.g., the cable or satellite service) would like to enable people to buy more than one pay-per-view service at a time, but this requires distributing the signal to more than one television at the same time. Because most homes are not wired for cable in all of the rooms where the televisions are located, the providers would like to enable the user to self-configure the network without incurring the expense of running cables. In addition, the optimal place for bringing in the signal, either from the curb for cable or from a wall or roof-mount antenna, is often not where the users want to watch television. The content providers incur high costs in running cables for the customers; yet, they have to do this or the customer will not buy the service. The concept of a "self-install" that does not require running cables has made the carriers very interested in wireless video distribution systems.

Even in the case of the home theater system, connecting the cables correctly to all of the components of the system can be quite daunting. Wireless connectivity with service discovery would allow people to bring in a new component and have it automatically find all of the other components and create the required connections. The user would only have to plug in power for the new component, and all of the components could self-configure the connections. For the typical consumer, this is a great advantage from the point of view of ease of use. Not only that, but it also will allow the user to position the components wherever they want.

Streaming multimedia is not the only type of application that 802.15.3 enables. Two-way multimedia applications, e.g., where the person interacts with the application, are also a good application for 802.15.3's QoS capability. If the application includes two-way voice interaction, then the application needs both the low latency for voice as well as the high bandwidth for video. Some examples of this kind of application are video conferencing, video intercoms, and gaming. However, it may be possible to separate the delivery of voice from video and to prioritize the voice delivery. People tend to be more accepting of glitches in the video presentation than in the audio portion. In this case, each DEV would allocate two streams, one for video and one for audio, to the other DEV. If the application supported scheduling audio and video together in a single stream, then the DEVs would require only one time allocation in the superframe.

Another application is a detachable display for a video camera. Sharp® demonstrated this application using a proprietary radio from their research departments that carried data at 10 Mb/s. An 802.15.3 radio would give greater data rate and lower cost because it was based on a standard. This application allows the user to put a camera on a tripod and take the display to a remote location. Because 802.15.3 enables two-way communications with low latency, the user would be able to control the camera from remote as well as be able to see the images that the camera was recording. Digital still cameras would also benefit by having a remote display with shutter release and focus control.

Sony™ has pioneered the concept of robots as entertainment devices, and this idea is beginning to take hold at other companies as well. One of the enhancements that has been suggested is remote control of the robot with a remote display. The user would be able to give voice commands to the robot and see what the robot is doing. Another application for remotely viewing with a camera would be for remote control vehicles like model cars and possibly model airplanes (although the range would have to be extended). This would allow the person controlling the vehicle to change their viewpoint to be from the inside of the vehicle looking out. It would also allow the vehicle to go behind visual obstructions while remaining under the control of the user or even to integrate the vehicle in a computer game.

Digital and analog camcorders are an important application for wireless connectivity. Although it is possible to view the video on the camera, the displays are usually small (less than 4 inches) because the camera needs to be small as well. Currently, users need to find the video cable for the camera and attach it to a television to show the video. Although some televisions now have connectors in front, not all do, and the cables can get lost. With a wireless link, users would be able to easily display the contents of their camcorder on any display device, e.g., television or computer monitor. Not only that, they could make copies of the video directly to tape or DVD without having to connect the cables to the back of the VCR or DVD recorder. If users had a personal video recorder (PVR), they would be able to use the camera for control, the television for display, and the PVR as an editing device. Once the edited video was complete, the PVR could record it to tape or DVD.

In summary, some of the possible multimedia applications are:

- Video distribution from set top boxes to televisions
- Wireless connection of home theater and stereo components
- Wireless video conferencing, video intercoms, and interactive gaming
- Wireless displays for entertainment applications like robots and remote-controlled vehicles
- Remote displays and controls for video and digital still cameras
- Displaying video from camcorders to televisions
- Storing video from camcorders to VCRs and DVD recorders
- Editing videos using a camcorder and PVR

Chapter 3 Overview of the IEEE 802.15.3 standard

Networks are an essential part of computers today and have become a key component to many of the applications that run on them. These networks may serve different geographic areas and subsequently have different implementations. The areas served by these networks can be divided into four types:

a) Wide area networks (WANs): A country or even the entire world

b) Metropolitan area networks (MANs): A city or similar area

c) Local area networks (LANs): A single building or campus

d) Personal area networks (PANs): An area around a person.

The last type of network, the PAN, is a relatively new area of development. In a PAN, the group of connected devices is not referred to as a network, but rather as a piconet, to imply that it is a small group of networked devices. PANs are designed to connect the many electronic devices that a single person might own to enable data sharing, connectivity, and new applications that are still developing. Because of this, PAN devices are used in a way that is very different from other networks.

Early definitions of a PAN, including the one used in the IEEE Std 802.15.3, referred to it as being limited in distance[12]. However, the results of the call for applications (CFA) included multiple applications that required relatively long distances, e.g., coverage of an entire house. Instead of range, one of the key characteristics of a PAN that differentiates it from a LAN is that a PAN normally consists of DEVs that are under the control, in some fashion, of a single user. A LAN, on the other hand, normally consists of devices that belong to multiple users. Thus the connectivity focus of a PAN tends to be inward, whereas a LAN's connectivity focus is outside of the LAN. The typical characteristics of a PAN are as follows:

---

[12] The definition used in the standard was "... a space about a person or object that typically extends up to 10 m in all directions..."

---

a) A single person owns all of the devices.

b) A PAN should form automatically and should exist only for as long as it is needed.

c) The connectivity within a PAN is peer-to-peer among the members of the PAN, although external connectivity is allowed.

In a typical PAN, a single user[13] owns or controls all of the devices. On the other hand, in a LAN, a typical user may not have even seen all of the devices, let alone own or control all of them. In the PAN environment, this allows the user to move the devices around for better reception and to be able to use different authentication processes to enable security. Another difference between PANs and LANs is that although LANs cover relatively small areas, the main use of LANs is to enable connectivity with WANs, in particular, with the Internet. Although a LAN can be valuable in providing connectivity in a building or campus, most people also use the LAN to bridge to the Internet. It is much more likely that a LAN device will be trying to reach other devices on the Internet than in a PAN, where it will be possible that the devices will not have a connection to a WAN.

Another use for a PAN, especially a WPAN, is for disposable networks. In this case, two devices that come in range are able to quickly establish a network to communicate data and just as quickly are able to take down the network when they are done. Fast connect times are important as well as the ability to spontaneously form a network without the intervention of a networking expert. A typical application involves a user walking up to another wireless-enabled device that is able to advertise its services and quickly fill the need. This might involve a digital still camera and a photo-processing kiosk. The user is able to select the number and type of prints on the camera and quickly download the pictures and the printing instructions to the kiosk for printing.

A PAN also has the characteristic that the devices will generally want to connect directly with the other devices in the PAN, i.e., in peer-to-peer mode. Traffic in the LAN, on the other hand, is often directed to devices that are not in that LAN, but rather is routed to other devices that are often quite distant. In the case of IEEE Std 802.11, stations in infrastructure mode are not allowed to communicate

---

[13] Or a small, closely-related group of people, e.g., a family.

directly with each other, but need to send all data to and receive data from the access point.

ELEMENTS OF THE 802.15.3 PICONET

There are two key elements to the 802.15.3 piconet: the piconet coordinator (PNC) and the device (DEV), as shown in Figure 3–1. The PNC is the center of the piconet, and the extent of the piconet is determined by the ability of DEVs to hear the beacon transmitted by the PNC. The PNC also acts as a DEV in the piconet and is able to communicate with any other DEV in the piconet. Every 802.15.3-compliant implementation is a DEV, but not all DEVs are capable of becoming a PNC. However, without a PNC, it is not possible to form an 802.15.3 piconet. This leaves open the possibility that two 802.15.3-compliant DEVs would be unable to communicate if neither one of them implemented PNC capabilities. Although this is a potential problem, the task group felt that for some very cost sensitive applications, like speakers, the PNC capability would be an unnecessary expense. In the case of a speaker, it does not have a function in the absence of a sound source, e.g., microphone, DVD player, MP3 player, component receiver, and so on; therefore, the assumption was that two speakers would not need to form a piconet but rather could rely on the sound source to take on the PNC role.

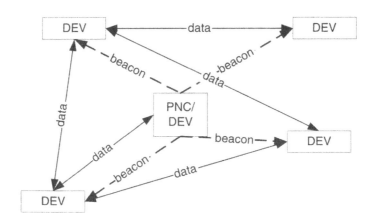

Figure 3–1:   Elements of an 802.15.3 piconet

The PNC is required in every 802.15.3 network because it operates as the arbitrator in assigning channel time. The PNC also provides the timing reference for the piconet via the beacon. All DEVs need the approval of the PNC to join the piconet via association, and if security is enabled in the piconet, they also need to authenticate to the PNC to gain membership in the piconet. The PNC also keeps track of sleeping DEVs to enable them to save power while making sure that their channel time is allocated when they are scheduled to be awake.

Although the PNC has a variety of responsibilities, the standard has made an effort to keep most of the complexity, whenever possible, in the DEV. The reason behind this is that it allows DEVs to be as complex as they need to be to support their application without burdening the DEV that happens to be the PNC. An example of this is channel time allocation. Rather than sending the required data rates, buffer sizes, latencies, and priority to the PNC so that it can determine the appropriate amount of channel time, the DEVs are required to calculate the amount of channel time that they require for their application and the frequency with which it needs to be allocated. The PNC simply decides if there is enough time in the superframe to allocate the channel time. In the case of an application specific DEV, e.g., a microphone or game controller, the DEV already knows the required channel time and allocation frequency and always makes the same channel time request.

The DEVs in the piconet can also take on special roles to fulfill other requirements. For example, a DEV in the piconet can request channel time from the PNC to become the PNC of another piconet, called a dependent piconet. A DEV that is the PNC of a dependent piconet is referred to as a dependent PNC. A dependent piconet shares channel time with the original piconet, called the parent piconet, as shown in Figure 3–2. If a DEV is a dependent PNC, it is acting in two separate roles: the first as a DEV in the parent piconet and the second as the PNC of another piconet that is synchronized to the parent piconet.

An example of a parent and dependent piconet is illustrated in Figure 3–3. In the figure, the dependent PNC is a member of both the parent and dependent piconets. Also note that one of the DEVs is a member of both the parent and the dependent piconet. The standard does not forbid a DEV from joining both piconets. In this case, the DEV would need to keep track of two separate personalities, one as a DEV in the parent piconet and the second as a DEV in the dependent piconet.

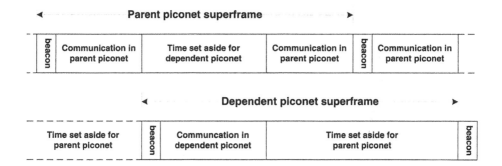

Figure 3–2: Timing relationship for parent and dependent piconets

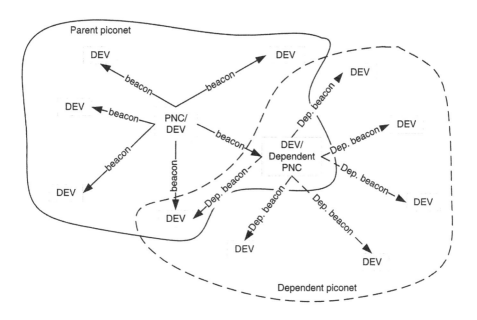

Figure 3–3: Possible connections with parent and dependent piconets

There are two types of dependent piconets—the child and the neighbor piconet. The difference between the two piconets is determined by the type of membership the dependent PNC has in the parent piconet. If the DEV is a full member of the parent piconet, then it can request channel time to form a child piconet and become a child PNC. A child piconet can be used to extend the range of the original piconet by using the child PNC as a bridge to other DEVs. A child piconet can also be used to provide different level or type security to a subset of the DEVs in the parent piconet.

On the other hand, if the DEV does not want to or is unable to become a full member of the parent piconet, it joins the network as a neighbor PNC. A neighbor PNC is not able to communicate directly with members of the parent piconet (other than with the parent PNC) and it is not given the parent piconet's group data encryption key. A neighbor piconet can be used to support a different level of data security or to share channel time between two piconets controlled by different users.

A DEV may also act as the key originator in the piconet. The key originator is a DEV that generates symmetric encryption keys to protect the confidentiality and integrity of data for subsets of the DEVs in the piconet. In a secure piconet, the PNC is the key originator for all of the DEVs that are members of the piconet. A DEV might take on the role of a key originator if it does not trust the PNC or if it wants to secure an application for a limited set of DEVs with a different security method than the one used by the piconet as a whole. In this role, the DEV authenticates peer DEVs in the piconet and maintains the management and symmetric keys for the DEVs. A key originator performs all of the security-related actions of the PNC with the exception of sending a secure beacon.

PHY OVERVIEW

The 802.15.3 PHY operates in the 2.4–2.4835 GHz frequency band that is available as an unlicensed band in most of the countries in the world. This allows a person to use their 802.15.3-compatible device in Europe, Japan, Canada, and many other countries. The symbol rate is 11 Mbaud with either 1, 2, 3, 4, or 5 bits per symbol to give 11, 22, 33, 44, and 55 Mb/s as a raw data rate.

The base rate for the 2.4 GHz PHY is 22 Mb/s. The base rate is used for the beacon, association, disassociation, broadcast, multicast, and whenever the DEV is unsure what rate is supported by another DEV or the PNC. All DEVs, therefore, are required to support the base rate, whereas the other data rates and modulations are optional. The 11 Mb/s data rate was added to the standard to enable communications between DEVs that are distant from each other in the piconet. For example, consider the piconet shown in Figure 3–4 with three DEVs and the PNC. As Figure 3–4 illustrates, communication in the piconet can take place directly between members of the piconet, e.g., between DEV-1 and DEV-3, and does not have to first go through the PNC. This is an important difference between 802.15.3 and 802.11's infrastructure mode, which requires that all traffic goes through the access point.

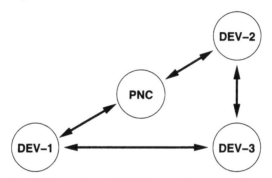

Figure 3–4: DEVs using 11 Mb/s mode to enable data transfer

If the distance between the PNC and DEVs 1–3 is near the limit of what can be reached with 22 Mb/s, the DEVs will be able to join and participate in the piconet. However, although the distance between DEV-2 and DEV-3 is short enough to enable peer-to-peer communication, the distance between DEV-1 and DEV-3 or DEV-1 and DEV-2 will be too long for reliable 22 Mb/s communications. In this case, the DEVs are still able to use the 11 Mb/s transfer mode to communicate between themselves.

The 802.15.3 PHY intentionally reused many of the characteristics of the 802.11 direct-sequence spread spectrum (DSSS) and 802.11b PHYs while seeking to relax the requirements wherever possible. For example, 802.15.3 uses the same frequency accuracy and roughly the same channel plan as 802.11b, but it has a relaxed adjacent and alternate channel requirements for the receiver to enable low-cost implementations.

The IEEE Std 802.15.3 Working Group selected a coding technique, trellis-coded modulation (TCM), that has been proven reliable and efficient in wireline standards for decades. This coding method requires a smaller signal-to-noise ratio for the demodulation of 11 Mb/s QPSK-TCM than what is required for the 11 Mb/s CCK mode of 802.11b. Thus, for the same transmit power and receiver noise figure, an 802.15.3 PHY will have a greater range at 11 Mb/s than will an 802.11b radio at 11 Mb/s.

The initial work for the 802.15.3 PHY was aimed at compliance with the FCC's "low power rules" (i.e., 47 CFR 15.249 [B29]) rather than the spread spectrum rules that were used for 802.11b (i.e., 47 CFR 15.247 [B29]). The low power rules would have limited 802.15.3 implementations to about +8 dBm of transmitter power. However, a recent FCC report and order changed the requirements for spread spectrum systems from a jamming resistance test to a power spectral density test. Although this change was intended to allow the proposed 802.11g PHYs to operate in the U.S., it also allowed 802.15.3 radios to use higher transmitter power as well. In the U.S., an 802.15.3 PHY can transmit at up to 1 W with up to 6 dB of antenna gain. In the European Community (EC), both 802.11g and 802.15.3 are limited to 100 mW, including any antenna gain. A typical implementation of 802.15.3, however, would use a transmitter power similar to the current range for 802.11b stations, i.e., about +12 to +18 dBm. However, an extremely low power implementation of 802.15.3 might use a lower transmit power, trading range for battery life. The standard does not have a minimum transmit power requirement.[14]

One of the key differences between 802.11b (or 802.11g) and 802.15.3, other than modulation techniques, is the channel spacing and transmit power spectral density. IEEE Std 802.11b defined a transmit power mask that, by today's standards, is very lax. With digital modulation and pulse shaping that is typically used in current radio implementations, it is relatively easy to control the transmitter spectrum to be less than the 22 MHz that is allowed for 802.11b implementations. Compliant 802.15.3 implementations restrict their transmit

---

[14] This is also true for all but two of the nine 802.11 PHYs. 802.11 FHSS and 802.11 DSSS do specify a minimum required transmit power.

bandwidth to less than 15 MHz. This allows up to four non-overlapping channels in the 2.4 GHz band instead of the three channels specified for 802.11b. The narrower transmit bandwidth reduces the interference to other networks, like 802.15.1 and 802.11b, whereas the narrower receiver bandwidth reduces the susceptibility of 802.15.3 to interference from other wireless networks or interference sources. The reduced susceptibility of 802.15.3 to other interferers is very important because the number of wireless devices in the market is increasing at an incredible rate. As the unlicensed frequency bands becomes more congested, the ability of 802.15.3 radios to reject interference and maintain a high data rate will become increasingly important.

There are four primary sources of interference in the 2.4 GHz band: 802.15.1 WPANs, 802.11b WLANs, microwave ovens, and cordless phones. The coexistence performance of 802.15.3 with 802.11b and 802.15.1 was analyzed and placed in Annex C (informative) of IEEE Std 802.15.3, and it is also discussed in Chapter 9 of this book. In general, 802.15.3 piconets can achieve an acceptable level of coexistence with other devices in the 2.4 GHz band. In addition, IEEE Std 802.15.3 and IEEE Std 802.15.2 provide methods that can enhance the coexistence of these networks. These methods are also discussed in Chapter 9. Cordless telephones in the 2.4 GHz band usually are narrowband frequency hopping devices, and so they would tend to affect 802.15.3 WPANs in a manner similar to 802.15.1 WPANs. Microwave ovens, on the other hand, are wideband interferers that operate periodically in the band. There are techniques, out of the scope of IEEE Std 802.15.3, that can be used to mitigate the effects of the microwave oven noise.

STARTING A PICONET

The first activity for an 802.15.3 piconet is its creation. A PNC-capable DEV, instructed by its higher layers, begins to search the channels to find other 802.15.3 piconets and to determine which channel has the least amount of interference. If the DEV is unable to find another 802.15.3 piconet, it would then start its own piconet by broadcasting a beacon. The initial draft for IEEE Std 802.15.3 included a selection process for the PNC wherein PNC-capable DEVs announced their capabilities so that the "most capable" DEV would be selected to be the PNC. However, after some consideration, this process was removed because it is unlikely to happen unless all of the DEVs begin scanning simultaneously. Even in the case where all of the DEVs are turned on at the same time, e.g., if there was a power outage, the DEVs will take different amounts of time to "boot up" before turning on the wireless interface.

Instead of a selection process, the first PNC simply looks at the capabilities of DEVs that associate with the piconet to see if a more capable DEV has joined. If the new DEV is better suited to become the PNC, the old PNC will normally hand over responsibility to the new PNC. Thus, in the typical case where the piconet grows slowly, the "best" DEV will eventually become the PNC without the complexity of the selection process. The handover process also allows a PNC that is turning off to pass over coordination responsibilities to another DEV that is remaining active in the piconet. The handover process passes the association information, DEV capabilities, channel time requests, security information and power management states of the DEVs to the new PNC. The intention is that the handover process will be seamless and will only minimally disrupt traffic in the piconet.

Although a network normally needs at least two devices to be considered an actual network, an 802.15.3 piconet is said to be in operation as long as the PNC is sending beacons, even if there are no other DEVs in the piconet. If the beacon from the PNC disappears, due to power outage, interference, etc., the piconet will cease to exist after the timeout specified in the standard. The remaining DEVs in the old piconet that are PNC capable have the option to start a new piconet using the regular piconet start procedure.

THE SUPERFRAME

The basic timing for the piconet is the superframe, as shown in Figure 3–5. The superframe is made up of three parts:

a) The beacon

b) The contention access period (CAP)

c) The channel time allocation period (CTAP)

The beacon is sent by the PNC at a regular interval such that the superframe duration is normally constant. The PNC can change the superframe duration, but it announces the change to the DEVs in the piconet via the beacon before it makes the change. The superframe duration is sent in each beacon so that all of the DEVs will know when the next beacon will occur. The beacon also indicates when the CAP ends as well as the channel time allocations (CTAs) in that superframe. The beacon is used to communicate management information as well as to indicate which DEVs are sleeping and when they will be awake.

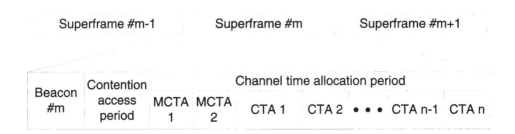

Figure 3–5:   The 802.15.3 superframe

The CAP is used to transfer commands and data either between the PNC and a DEV or between DEVs in the piconet. The access method for the CAP is carrier sense multiple access with collision avoidance (CSMA/CA) in a manner similar to IEEE Std 802.11. This access method allows a very efficient use of the time in the CAP while providing a low latency for DEVs that wish to send commands or data. The CAP is well-suited to sending small amounts of asynchronous data between DEVs in the piconet because it does not require a formal channel time reservation process. The CAP is also used by DEVs that are associating with the piconet to send the Association Request commands to the PNC. The length of the CAP may vary between superframes, and the PNC may even set it to be a zero length for some of the superframes. The PNC is allowed by the MAC standard to restrict the type of commands or data that may be sent in the CAP, but for the 2.4 GHz PHY, the PNC is required to allow all DEVs to use the CAP for commands, association, or data.

The CTAP is composed of one or more CTAs that are allocated by the PNC to a pair of DEV identifiers (DEVIDs). Only the DEV that is listed as the source is allowed to transmit in a CTA, and so the access method for CTAs is time division, multiple access (TDMA). In the case where the CTA has the PNC identifier (PNCID) as either the source ID (SrcID) or destination ID (DestID), the CTA is called a management CTA (MCTA). In most cases, there is no difference between an MCTA and a CTA. The only exceptions to this are when the source address is either the unassociated ID (UnassocID) or the broadcast ID (BcstID). In this case, there are multiple DEVs that are allowed to send information in the MCTA. To handle potential collisions, a third type of access method, called slotted aloha, is used. Although this method is generally not as efficient as using the CAP for association, commands, and data, it was put in to allow for future PHYs that may not be able to perform clear channel assessment (CCA) efficiently.

Using MCTAs instead of the CAP also adds complexity to the PNC because it now needs to keep track of the potential needs of all the DEVs in the piconet and assign MCTAs periodically to all of them. The DEVs need periodic MCTAs so that they can send commands to the PNC and maintain their association in the piconet. When the PNC allocates the CAP, it does not have go through the effort to separately allocate time for association, commands, and periodically

> The unassocID was originally called the AssocID because it was used by the DEV to associate with the piconet. The name was changed to be the UnassocID in the letter ballot because the only DEVs that use this DEVID are unassociated ones. On the other hand, MCTAs that are reserved for the association process are referred to as association MCTAs.

directed links to the DEVs in the piconet. The use of open (BcstID as the SrcID) or association (UnassocID as the SrcID) MCTAs is optional for the protocol and for the 2.4 GHz PHY. Because there is no difference between a regular MCTA and a CTA, all DEVs are required to support the TDMA access method.

JOINING AND LEAVING A PICONET

Once the piconet is established by the PNC, DEVs begin to join the piconet using the association process. The association process is used to limit the size of the piconet so that the PNC is not overloaded and to assign an ID to the incoming DEV. Rather than sending the complete MAC address, which is 8 octets, for the source and destination in the MAC header, IEEE Std 802.15.3 requires that new DEVs joining the piconet get an 8-bit ID, the DEVID, from the PNC to identify the DEVs messages in the piconet. This choice cuts 14 bytes of overhead from the MAC header, which improves the efficiency of the 802.15.3 protocol. The PNC is able to refuse an association request for many reasons, including the reason that the PNC does not have the resources to handle another DEV in the piconet.

Once the DEV has joined the piconet, it ceases to be a member of the piconet either because it requests to leave, the PNC tells it to leave, or through inactivity in the piconet. When the DEV wants to leave the piconet, it sends a Disassociation command to the PNC. Likewise, if the PNC wants the DEV to leave the piconet, it sends a Disassociation command to the DEV to tell the DEV that it is no longer a part of the piconet. When the DEV joins the piconet, the PNC assigns the DEV an association timeout period (ATP). If the DEV does not successfully send a frame to the PNC at least once during an ATP, the PNC will consider the DEV to be disassociated and will eventually reuse the DEVID it assigned to that DEV for another DEV. In a similar manner, if the DEV does not hear any communication from the PNC for an ATP, it considers itself disassociated and needs to search for a piconet to join.

CONNECTING WITH OTHER DEVICES

Once a DEV has joined the piconet, it finds out about other DEVs in the piconet when the PNC broadcasts the PNC Information command. This command contains a list of all the DEVs in the piconet, their MAC addresses, and their capabilities. With this information, the DEV is able to send small amounts of data to the other DEVs during the CAP or to request channel time from the PNC to send frames to other DEVs.

There are two types of channel time requests (CTRqs) that the DEV may make to the PNC. The first type of CTRq, called an isochronous CTRq, is a request for channel time on a periodic basis that does not expire. Until the request is terminated by either the source, destination, or PNC, the CTRq remains valid. If the PNC agrees to the request, it issues a stream index that is unique in the current piconet and that corresponds to a specific source and destination pair of DEVs. A DEV is able to request more than one allocation per superframe (a super-rate allocation), one allocation per superframe, or one allocation every *n* superframes (a sub-rate allocation).

The different types of isochronous allocations are used so that the DEV can request the latency and throughput that is consistent with its application.

Insufficient bandwidth or too much latency can make or break a product. Thus, the ability to have a DEV request and receive a specific amount of latency and throughput from the network is extremely important. This ability alone will open several new market segments for 802.15.3 solutions.

The second type of CTRq is called an asynchronous CTRq. This CTRq is a request for a total amount of channel time that will be allocated whenever the PNC is able to fit it into the superframe. Because it is a request for a total amount of time, once the PNC has allocated all of the requested time, it drops the asynchronous CTRq and no longer attempts to allocate the CTAs. This method is slightly more efficient than using an isochronous CTRq because the DEV does not need to explicitly terminate the CTA when it is done sending data. An asynchronous allocation is not assigned a stream index by the PNC. Instead, all DEVs use the same stream index for asynchronous data.

For both asynchronous and isochronous CTAs, the source may request a modification of the current allocation using the CTRq command. As with the initial allocation, the PNC is able to grant or refuse the request depending on the capabilities of the PNC and the current traffic conditions in the channel. A request to modify an asynchronous allocation results in an update to the amount of time that the DEV is requesting the PNC to allocate.

The source, destination, or PNC may terminate the existing isochronous allocation. The PNC always grants the termination request if all of the following are true:

a) the allocation exists,

b) the DEV that sent the request is a valid member of the piconet, and

c) the requesting DEV is either the source or destination of the allocation.

The source DEV or the PNC may terminate an asynchronous allocation using the CTRq command. Because the asynchronous allocation is somewhat transient, the destination DEV is not allowed to request a termination of an asynchronous CTA.

DEPENDENT PICONETS

Dependent piconets are WPANs that rely on the parent piconet for their channel time and synchronization. The basic idea is that the two piconets are synchronized in time and set aside time in their allocations to allow quiet time for the other piconet to operate.

One type of a dependent piconet is the child piconet. The PNC of the child piconet is a full member of the parent piconet and so it is able to communicate with other DEVs in the parent piconet as well as with DEVs in the child piconet. Thus, the child PNC is able to act as a bridge between the two piconets. Because the child PNC is a full member of the parent piconet, it must be a fully compliant 802.15.3 DEV, as well

> The child piconet was originally called the daughter network, but it was renamed to be somewhat more politically correct. Task group discussion of creating dependents for a parent, the responsibilities of the parent for the child, or what happens with an ill-behaved child always seemed to evoke side jokes about how this might apply to real life. Because of the wide variety of anthropomorphizing available for parents, children, dependents, and neighbors, the discussion of dependent piconets was often one of the more humorous ones in the standards development process.

as being PNC capable. Some of the applications for child piconets are:

- Extending the range of the parent piconet by using the child PNC as a relay to DEVs that the child PNC is able to reach;

- Setting up a separate group for security purposes if the child PNC does not trust the parent PNC;

- Allowing different piconets to share the same channel without interference when there are no other channels available;

- Allowing the formation of dedicated sub-networks that need to control their time allocations separately (e.g., a set of speakers and a component amplifier that are pre-configured as a network at the time of manufacture).

The first step in becoming a child PNC is for the DEV to join the parent piconet using the association process. Once the DEV has joined the parent piconet, it requests channel time using the CTRq command with its own DEVID as both the source and destination. If it is granted time by the PNC, the child PNC can then begin transmitting its beacon in the allocated time. The child PNC sets aside time in its superframe that corresponds to the time that the parent piconet is operating.

The other type of dependent piconet is the neighbor piconet. The key difference between the neighbor PNC and the child PNC is that the neighbor PNC is not a full member of the parent piconet. In fact, the neighbor PNC is not even required to be a fully compliant 802.15.3 DEV. The

> The name "neighbor" was meant to indicate that the parent PNC would be willing to give the DEV some time, but that it was not really part of the parent's piconet. This is like loaning lawn tools to the neighbors, but not making them a part of the family like your children. See, dependent piconets are one of the funniest parts of the standard as well as being quite useful.

neighbor PNC is only required to support a subset of the 802.15.3 commands, and it only needs to support the base rate. One of the unique applications of the neighbor piconet is that it allows controllers from other standards to implement a method of coexistence in which the two networks can negotiate to share time in the same channel without causing collisions. As an example, if an 802.11 AP implemented a few 802.15.3 commands and modified the modem slightly (the PHYs are essentially the same for the base rate and the DSSS mode), it would be able to request time from an 802.15.3 PNC so that it could schedule the traffic for the WLAN when the WPAN was quiet. Similarly, the 802.15.3 PNC would use the quiet time in the WLAN to operate the WPAN.

The neighbor piconet can also be used if the PNC-capable DEV is unable to become a secure member of the parent piconet. It would then be able to set up a separate piconet with different security while sharing the same channel. Note that the neighbor PNC is not able to communicate with the DEVs in the parent piconet (unless they also join the neighbor piconet). The key difference in forming a neighbor piconet is that the DEV associates with the parent piconet not as a regular DEV, but as a neighbor PNC. If the parent PNC allows the formation of a neighbor piconets, it will assign the neighbor PNC a special DEVID, the neighbor ID (NbrID), that indicates that it is a neighbor PNC and not a member of the piconet. Once the neighbor PNC has joined the piconet, it can then request channel time for its piconet in the same way that the child PNC requests channel time.

The parent PNC is allowed to force the dependent piconet to stop operations whenever it wants for any reason. For a child piconet, it only terminates the CTA allocated for the child piconet because the child PNC may want to continue as a DEV in the parent piconet. In the case of the neighbor piconet, the PNC uses the

Disassociate command to cause the neighbor PNC to cease operations. In this case, there is no reason for the neighbor PNC to remain in the parent piconet if it is not going to be given channel time for the neighbor piconet. In all cases, the PNC will listen for the child or neighbor PNC's beacons to allow it time to shutdown the dependent piconet. However, the PNC will eventually remove the allocation even if the child or neighbor has not been able to complete its shutdown process.

OBTAINING INFORMATION

IEEE Std 802.15.3 provides a variety of commands that are used to get information about other DEVs in the piconet or about the current channel conditions. The Probe Request and Probe Response commands are used to exchange information elements (IEs) with other DEVs. On the other hand, the Announce command is used to send, unrequested, an IE to another DEV in the piconet.

The PNC Information Request and PNC Information commands are used to request and provide information about the DEVs that are currently members in the piconet. The PNC Information command contains information about the DEV's capabilities, its MAC address, its DEVID, and other characteristics that are unique to the DEV. For example, other DEVs in the piconet can use the capability information to determine the data rates that are supported by the DEV for communication in the piconet.

The Remote Scan and Channel Status commands are used to determine information about the channel quality in one of the supported PHY channels. Because 802.15.3 operates in the unlicensed band, the DEVs are required to accept interference from primary and secondary users in the band. In addition, the DEVs could encounter interference from other 802.15.1 or 802.15.3 WPANs, 802.11 WLANs, microwave ovens, cordless phones, and so on. A DEV can get information about the quality of a link to a specific DEV with which it is communicating using the Channel Status Request command. The response to this command, the Channel Status Response command, contains information that can be used to determine the frame error ratio (FER) for that link. The PNC is also able to send a request for the collective channel status information to all of the

DEVs in the piconet by sending the Channel Status Request command as a broadcast frame.

Furthermore, the PNC can request that another DEV in the piconet scan a set of channels to determine if other 802.15.3 piconets are present as well as the channel that appears to be the best from an interference point of view. This information, together with its own assessment of the available PHY channels, can be used by the PNC to move the piconet to another channel.

POWER MANAGEMENT

One of the key advantages of IEEE Std 802.15.3 is its support of power management modes. The goal in any power management strategy is for the DEV to turn off as many of its operations as possible for the longest period of time. When the DEV is turned off or "sleeping," it will not be listening to beacons or other traffic. A DEV is "woken up" when it begins listening to beacons or time allocations on a regular basis. However, it is not enough for the DEV to "disappear" for some period of time. To maintain communications with a sleeping DEV, the other DEVs in the piconet will need to know when it will be awake again. These DEVs also need a method to wake up the sleeping DEV if traffic needs to be sent to it. In addition, if more than one DEV is sleeping, then the DEVs that are running the same application need to wake up at the same time to communicate.

To do this, the standard provides three power-save modes, asynchronous power save (APS), device synchronized power save (DSPS), and piconet synchronized power save (PSPS). A DEV that is not in a power-save mode and is therefore listening to every beacon is in ACTIVE mode. In any of the four power management modes, ACTIVE or the three power-save modes, the DEV may be in SLEEP state or AWAKE state, depending on the traffic in the piconet. If the DEV sees that it is not the source or destination of a particular CTA, it is not required to have its receiver active for that duration. This allows DEVs, even those in ACTIVE mode, to save power. The various groupings of power-related modes in the standard are listed in Table 3–1.

Table 3–1: Power-related modes in 802.15.3

Name of mode grouping	Valid modes
Power management modes	ACTIVE, APS, DSPS, PSPS
Power save modes	APS, DSPS, PSPS
Synchronized power-save modes	DSPS, PSPS

For greater power savings, the DEV may choose to switch to one of the three power-save modes. The APS mode offers the best power savings by giving up synchronization with the rest of the piconet. In APS mode, the DEV simply stops listening to beacons. All of the streams with the DEV as the source or destination are terminated by the PNC. The only responsibility that a DEV in APS mode has is to send a frame to the PNC at least once every association timeout period (ATP) in order to maintain its status as being associated

> The ATP is used by the PNC as a heartbeat for each of the associated DEVs. If the PNC does not hear from a DEV at least once every ATP, it assumes that the DEV has disappeared and the PNC disassociates the DEV. Likewise, if a DEV doesn't hear from the PNC within an ATP, it assumes that it is no longer a part of the piconet and that it has been disassociated.

with the piconet. When another DEV wants to request channel time with a DEV in APS mode, it sends a CTRq command the PNC. The PNC then sets the DEV's bit in the Pending Channel Time Map (PCTM) IE to tell the DEV that there is a CTA pending for the DEV. The sleeping DEV has the option to terminate the newly allocated stream if it does not want to wake up.

The next-best power savings method is the DSPS mode. In this mode, the DEVs wake up periodically to listen to beacons. A beacon during which the DEV is in AWAKE state is referred to as a wake beacon. The time period between wake beacons is called the Wake Beacon Interval. The beacon number when the DEV will be awake is called the Next Wake Beacon. A group of DEVs that share the same Wake Beacon Interval and Next Wake Beacon is called a DSPS set. Multiple DEVs can join the same DSPS set, which allows them to be awake at the same time. Because an 802.15.3 communications link is peer-to-peer, it is important that both the source and destination are awake at the same time so that they can exchange frames. DSPS mode allows DEVs with similar application requirements to be asleep for long periods of time while maintaining synchronization of their wake beacons. Furthermore, the DEVs can allocate sub-rate channel time, i.e., channel time every N beacons, aligned with their wake beacons. The only requirement is that the sub-rate interval is greater than the Wake Beacon Interval so that the sleeping DEVs are not required to wake up more often than their normal wake cycle. Figure 3–6 shows the wake beacons and sub-rate allocation for a DSPS DEV that has a Wake Beacon Interval of 4 (i.e., it is awake every 4th beacon) and has sub-rate allocation every 8th beacon (i.e., its CTA rate factor is 8).

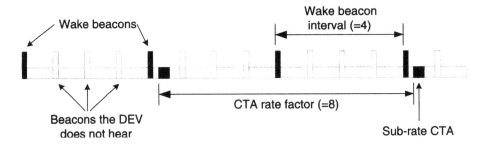

Figure 3–6: Wake beacons and Wake Beacon Interval for a DSPS DEV

The DSPS process begins when a DEV uses the SPS Configuration Request command to request that the PNC create a DSPS set with a specified interval. The PNC cannot change the interval; it must either reject or accept the request. The PNC is able to set the Next Awake Beacon in order to spread out the traffic in the piconet. The PNC is also able to

> All of the DEVs that belong to a particular DSPS set do not have to be in the same power management mode. Some of the DEVs in the set may be in ACTIVE mode while other DEVs in the set may be DSPS, PSPS or even APS modes.

change the Next Awake Beacon occasionally as the channel time requirements in the piconet change. Once the DSPS set has been created, it is assigned a DSPS set index to identify it. Other DEVs are then able to join this set by sending an SPS Configuration Request command to the PNC. A DEV may be a member of a DSPS set while it is in ACTIVE mode. The DEV then switches between ACTIVE and DSPS modes by using the PM Mode Change command. When the DEV is in DSPS mode, the PNC schedules asynchronous allocations for the DSPS DEV during its wake beacons. However, broadcast and multicast traffic is not necessarily aligned to the DSPS DEV's awake beacon because there can be more than one DSPS set, each with different intervals and Next Wake Beacons. If a DEV wants to send broadcast traffic to DSPS DEVs, it needs to send the frames during that DEV's wake beacons. The PNC must support a minimum of one DSPS sets, but it may support more depending on the implementation.

The last power-save mode is piconet synchronized power save (PSPS) mode. The primary goal of PSPS mode is not power savings, but rather ensuring that DEVs are awake for system changes and announcements. The DEVs in PSPS mode wake up at a time specified in the PS Status IE for the PSPS set (i.e., set 1). However, unlike DSPS modes, the DEVs in PSPS mode are not allowed to specify the interval between wake beacons. Instead the PNC chooses this interval based on requests from the DEVs in PSPS mode. The beacon when DEVs in PSPS mode are required to be in AWAKE state is called the system wake beacon. Because there is only one PSPS set for the entire piconet, the sleep interval, and hence the power savings, is a compromise between the requests of the DEV and the PNC's desire to change piconet parameters quickly. The PNC is allowed to change the interval between every system awake beacon. The PNC is prohibited from making major piconet changes until at least one system wake beacon has passed. This is intended to improve the chances that a DEV will hear the system changes before they happen. Of course, if any DEV misses the system change announcements, it will be able to recover from the changes by scanning for its piconet.

DEVs in PSPS mode keep their allocations when they switch into PSPS mode. However, because the system wake beacon may not be aligned to the PSPS DEV's sub-rate allocations, the PSPS DEVs need to wake up for not only the system wake beacons, but also superframes that contain its sub-rate allocations. Note that a DEV in PSPS mode with a CTA in every beacon is essentially in ACTIVE mode, because ACTIVE mode is defined as a DEV that listens to every beacon.

SYSTEM CHANGES

Because the PNC coordinates the activities of the piconet, it is also responsible for changing the parameters of the piconet whenever it is required. In particular, the PNC has the responsibility of choosing the PHY channel for the piconet and changing it whenever the channel conditions deteriorate. The PNC also sets and changes the superframe duration and is able to move the location of the start of the superframe in time. The PNC also selects the piconet identifier (PNID) and the beacon source identifier (BSID), which are used by the DEVs to identify the piconet.

Whenever the PNC wants to change one of these key parameters of the piconet, it puts the Piconet Parameter Change IE in the beacon to warn the DEVs of the impending change and to indicate the beacon number when the change will occur. The PNC will keep this IE in the beacon until the change occurs. In addi-

> The BSID acronym was based on the initials of one of the major contributors to the 802.15.3 standard. The acronym was selected first and then words were found that fit the initials. While not quite as funny as anthropomorphizing dependent piconets, it did seem humorous at the time.

tion, if there is a DEV in PSPS mode, the PNC needs to ensure that this IE is placed in at least one system wake beacon before the change takes place.

IMPLEMENTATION COST AND COMPLEXITY

One of the most important goals in the development of IEEE Std 802.15.3 was to create a standard for which low-cost implementations could be built. For the MAC, the cost and complexity of the standard was reduced in the following ways:

- Not all DEVs are required to be PNC capable. This allows very low cost implementations (e.g., a wireless speaker or remote control).

- The number of distinct frames, commands, and information elements was kept to a minimum.

- A DEV only needs to support the power-save modes that it needs. If the implementation is not battery-powered, the DEV would not need to support any power-save modes.

- The stream index for isochronous connections allows the DEV to easily separate data based on priority, both for transmission and reception.

- The fragmentation numbering was selected to allow the DEV to easily defragment the received frame with a minimum amount of processing.

- A simple PNC can be built that only uses a very rudimentary channel allocation algorithm. This is useful in applications such as personal digital assistants (PDAs) which are cost sensitive, but require the ability to run a simple piconet for data transfer (e.g., PDA to PDA transfer).

• The PNC is not required to forward data between DEVs. Unlike the 802.11 AP, which is required to store and forward data for power save STAs, the 802.15.3 PNC only keeps track of when a DEV will next be awake. The sending DEV can use this information to determine when it will be able to successfully send data to the power save DEV.

The 802.15.3 PHY was also selected to allow for very low cost implementations. Some of the design decisions that allow for cost reductions in some implementations are:

• A DEV is only required to implement the DQPSK mode, which does not require coherent reception. In addition, it can be received even when there is clipping in the receiver chain.

• The transmit power can be selected for the application. For some applications, a power amplifier (PA) will not be necessary and so the cost can be reduced.

• No special coding is required for the base mode (22 Mb/s DQPSK)

• The frequency and timing accuracies (+/- 25 ppm) as well as the symbol rate are the same as in 802.11b. This allows the implementer to re-use crystals that are now in relatively high volume production for 802.11b and 802.11g.

• Standard 802.11b RFICs can be used for the 22 Mb/s mode while 802.11g RFICs can be used for the higher order modulations. The design work has already been completed for these ICs and they are in relatively large volume production, reducing risk and price.

• The required sensitivity for the standard was relaxed to allow lower-cost radios.

• The frequency range (2.4 to 2.5 GHz) allows RFIC designers to use low-cost SiGe or RF-CMOS processes for the radio design.

• The PHY was designed to allow direct-conversion receivers, which can greatly reduce the cost and area of an implementation.

Chapter 4 MAC functionality

MAC TERMINOLOGY IN IEEE STD 802.15.3

One of the tasks in writing a good standard is to choose a set of terminology and to consistently use it. For example, in IEEE Std 802.15.3, the basic data structure is called a frame instead of a packet. Although the term "packet" would have been acceptable and is used in other documents to indicate essentially the same thing, the technical editor, with input from the Task Group, decided instead to use the term "frame." There are five types of frames in the standard: beacon, immediate acknowledgment (Imm-ACK), delayed acknowledgement (Dly-ACK), command, and data. "Command frames," or simply "commands," are used to communicate control information between the DEVs and the PNC or among DEVs. The standard also uses information elements (IEs) that are a type, length, and value encoding of information. IEs are used in cases when there is a variable number of elements that need to be transferred among the PNC and DEVs. The standard also adopted the use of the word "octet" instead of "byte" when referring to 8 bits of data.

Probably the most obvious difference in terminology between IEEE Std 802.15.3 and IEEE Std 802.11 is that members of the 802.11 WLAN are called stations (STAs), whereas in 802.15.3, they are called devices (DEVs). In 802.11, the STAs are coordinated, in infrastructure mode, by the access point (AP), whereas in 802.15.3, the DEVs rely on the piconet coordinator (PNC) to maintain order in the piconet. The decision to use different terminology was to draw a distinction between the architecture of the two standards. In 802.11, the key purpose of an AP is to provide a link to another network, either wired or wireless, that expands the reach of the WLAN. Often, this is a link to a wired network that then has access to the WAN, particularly to the Internet. On the other hand, in 802.15.3, the PNC is not required to have access to any other network. The PNC is used, not for access to other networks, but rather to control access to the time resources of the piconet. The piconet may have a DEV that acts as an access point, but that DEV is not required to be the PNC of the piconet. In addition, the PNC is also a DEV in the piconet and so it is just as likely to be a source or sink for a data stream as any other DEV in the piconet. An 802.11 AP, on the other hand, usually functions only as the access point, it does not act as an endpoint for the data traffic.

The 802.15.3 standard differentiates among four different types of piconets that can be formed: independent, parent, child, and neighbor. In addition, child and neighbor piconets are also referred to collectively as dependent piconets when the text applies equally to both types of a piconet. An independent piconet is an 802.15.3-compliant piconet that is not a dependent of a parent piconet and does not have any dependent piconets as a part of its piconet. The PNC of the independent piconet does not have to synchronize with any other PNC, and it is not required to provide consideration for any other dependent PNCs. A parent piconet is one in which a dependent piconet has been given a channel time allocation (CTA). The PNC of the parent piconet, referred to as the parent PNC, does not have to synchronize with any other PNC. However, it is required to take additional action when it stops a dependent piconet (see "Parent PNC stopping a dependent piconet" on page 176).

The child piconet relies on the parent PNC for its channel time and it is synchronized to the beacon of the parent PNC. The PNC of the child piconet, called the child PNC, is a full member of the parent piconet and is allowed to communicate with any of the other DEVs in the parent piconet.

The neighbor piconet is like the child piconet in that it relies on the parent PNC for its channel time and it is synchronized to the beacon of the parent PNC. However, the PNC of the neighbor piconet, referred to as the neighbor PNC, is not a full member of the parent piconet. The neighbor PNC is not allowed to communicate directly with other DEVs in the parent piconet. The neighbor PNC is not even required to be a PNC-capable, fully-compliant 802.15.3 DEV; it could be an entity that is compliant to another standard and that implements the minimal amount of functionality required to create a neighbor piconet within an existing 802.15.3 piconet. Dependent piconets are described in more detail in Chapter 5.

A frame exchange sequence involves at least two DEVs. A DEV that is sending a frame is the source of the frame, whereas the DEV (or DEVs) that are the intended recipient of the frame is the destination. In a sequence of frames, e.g., data frame followed by Imm-ACK, the DEV that begins the sequence is referred to as the originator of the exchange.

The other DEV (or DEVs) involved in the exchange are referred to as the target of the exchange. As an example, consider a data frame/Imm-ACK exchange with two DEVs, DEV-1 and DEV-2. DEV-1 sends the data frame to the DEV-2 requesting an Imm-ACK. Thus, DEV-1 is the originator of the sequence and DEV-2 is the target for

> The key concept in understanding originator, target, source and destination is that the originator and target DEVs are unchanged during a frame exchange sequence, whereas the source and destination can change with every frame that is sent.

the duration of the exchange. The source of the data frame is DEV-1, and its destination is DEV-2. When DEV-2 sends the Imm-ACK frame, it is the source of the Imm-ACK frame and DEV-2 is the destination. However, DEV-1, the destination of the Imm-ACK frame, is still the originator and DEV-2, the source of the Imm-ACK frame, is the target.

FRAME FORMATS

The goal of a standard is to enable interoperability between devices built by different manufacturers. One of the key clauses (IEEE Standards terminology for a chapter in a standard) in the standard for interoperability is Clause 7, which defines the frame formats. Although it is extremely important that implementers agree in every detail of the frame formats, the actual formats are somewhat arbitrary. Thus, although Clause 7 of the standard is key in the actual implementation, the details of it are not very interesting from the point of view of understanding how the standard works. There are a few areas in Clause 7, however, that do require some special attention.

The first area of interest is the bit ordering for the frames. There are two types of ordering defined in the beginning of the clause; the order in which the bits, octets, and fields are transmitted over the medium and the order that they are represented in the standard. In both cases, there is no correct way to order the bits, octets, and fields; rather, there are some minor trade-offs in choosing one method over the other. The frame formats clause in the initial draft was based very closely on IEEE Std 802.11-1999 and directly copied some of the wording. In particular, most of the text on the bit ordering over the air is exactly the same as the text in IEEE Std 802.11. Thus, in IEEE Std 802.15.3, the least significant bits of the least significant octet, e.g., bit 0 of octet 0, is sent first, whereas bit 7 of octet N (of an N octet frame) is sent last over the air, as shown in Figure 4–1.

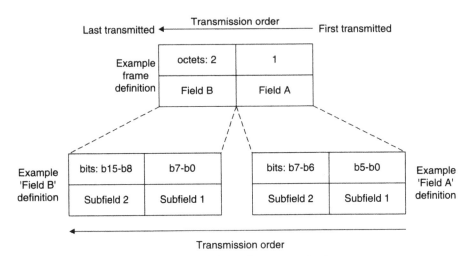

Figure 4–1: Bit and octet ordering

Despite the fact that the wording was taken directly from 802.11, which had multiple vendors providing interoperable solutions, some members felt that the text was a little confusing. In response to this, additional text and figures were added to clarify the bit and octet ordering.

One of the questions regarding bit ordering had to do with strings, like the one in the BSID IE. Suppose that users wanted to call their network "Wireless." Would the BSID IE be sent with "Wireless" or as "sseleriW;" i.e., would "W" or "s" be sent first? The draft indicates that the lowest order octet is always transmitted first. In the C language (and in many other programming languages), the first character in a string is indexed by the number zero. Thus, the lowest-order octet in "Wireless" is the ASCII encoding of "W," and so the string is sent as "Wireless," with the encoding of "W" sent first over the air. An example of the order that the data payload or a text field is transmitted is shown in Figure 4–2.

The other issue with bit ordering had to do with the representation of the frame formats in the draft. This mainly affected Clause 7, but it also affected a few figures and tables in the 2.4 GHz PHY clause, Clause 11. In this case, there is no technical difference if the figures have the first bits and octets on either the left or

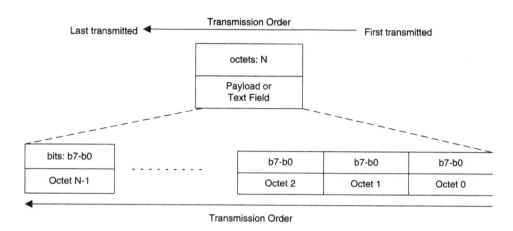

Figure 4–2: Example of payload or text field transmission order

the right side of the page. The key was to be consistent throughout the draft to minimize confusion. The issue is that numbers are usually represented with the most significant digits, octets, or bits on the left. However, many people are used to seeing the first bit sent on the left. Because the first bit over the air is the least significant bit, it was not possible to simultaneously satisfy both concerns. The decision was made to put the first bits, octets, and fields transmitted over the air on the right-hand side of the page so that the frame is transmitted from right to left.

The task group also attempted to minimize the number of frames, commands, and information elements. At the same time, the group also tried to avoid overloading any single command with too many functions. As with many things in the standard, however, it was not possible to simultaneously achieve both goals. In general, the group would err on the side of overloading the use of an existing command or IE rather than creating a new command or IE.

Piconet timing and superframe structure

Timing in the piconet is based on the superframe. The superframe begins with the first symbol in the PHY preamble of the beacon and ends a superframe duration later. The DEVs are required to set their timers to zero at the start of the PHY preamble of the beacon. Because the DEV will not know that the beacon has arrived until the MAC header is processed, the DEV never really sets its internal timer to zero; rather, it sets it to a time that corresponds to the time of the event when it knows the beacon has arrived. For example, if the PHY reports when the HCS has completed, the DEV would set its timer to the sum of the duration of the PHY preamble, PHY header, MAC header and HCS. The beginning of the PHY preamble is used because it represents the first transmission in the medium during the superframe. Thus, the last transmission during the superframe needs to end prior to the superframe duration. The timing for the piconet is illustrated in Figure 4–3.

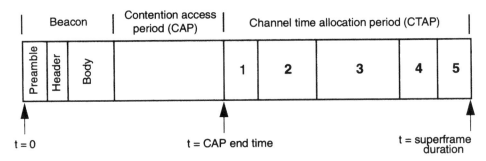

Figure 4–3: Piconet timing relative to the beacon

The value of the superframe duration is contained in the beacon and is generally constant during piconet operations. This allows DEVs that are sleeping to predict the timing of their wake beacon will occur so that they can wake up on time. The PNC is able to change the superframe duration, but it is required to warn the DEVs in the piconet via the Piconet Parameter Change IE in the beacon before it changes the value. This process is described in more depth in "Moving the beacon or changing the superframe duration" on page 136.

The beacon determines the start of the superframe and carries critical information regarding operation in the superframe. If a DEV misses the beacon, it will not be able to transmit unless it has been assigned a pseudo-static CTA (see "Channel time allocation period (CTAP)" on page 56). The beacon is used to communicate quite a bit of information to the DEVs in the piconet, and so it may grow to be a fairly large frame. As the number of octets in the beacon increases, the probability that it will be correctly received decreases. In addition, it may not be possible to fit all of the required information elements in the beacon due to PHY limitations on maximum frame size.

IEEE Std 802.15.3 overcomes this limitation by allowing the PNC to split the beacon into the beacon frame and one or more broadcast Announce commands that follow the beacon. The broadcast Announce commands that follow the beacon frame are referred to as the beacon extension and the combination of the beacon frame and the beacon extension is simply referred to as the beacon. The PNC will always send the CTA IEs, the BSID IE, and the Parent Piconet IE, if present, in the beacon frame. These IEs are the most critical to the operation of the piconet, and so they must occur in the beacon frame. In addition, the CTAs are all referenced to the start of the beacon frame, so if a DEV did not receive the beacon frame, it would not be able to use the allocations because it would not know when the superframe started. Other IEs that the PNC needs to send can be placed in the broadcast Announce commands that follow the beacon. The PNC indicates that the Announce commands will be following the beacon by setting the More Data bit to 1 in the beacon and in all but the last broadcast Announce command. Because the Announce commands in the beacon extension are sent to the broadcast address, the ACK Policy field is set to no-ACK.

The rest of the superframe is made up of the contention access period (CAP) and the channel time allocation period (CTAP), as shown in Figure 4–3.

Interframe spacings

IEEE Std 802.15.3 defines four different timings that govern the interframe spacing (IFS) in the superframe. All of these timings are PHY dependent because they are determined by the specific characteristics of the PHY. The

> In one of the proposals for the TG3a PHY [B5], the throughput improvement from using MIFS with Dly-ACK (5 frame bursts) was 11% at 110 Mb/s, 18% at 200 Mb/s, and 33% at 480 Mb/s for 1024 octet frames.

shortest of these IFSs is called the minimum interframe space (MIFS). This IFS is the amount of time that the PHY needs either between successive transmissions or successive receptions. This timing is only used in directed CTAs when the DEV that is sending the frame does not expect a response from the receiving DEV, i.e., when the source DEV either requests no ACK or is in the middle of a Dly-ACK burst. The MIFS was introduced to reduce the overhead and improve the efficiency for very fast PHYs.

The next shortest time is called the short interframe space (SIFS). The SIFS is the length of time required for the PHY to switch between transmit and receive. The SIFS is used when the DEV that is transmitting expects a response from another DEV, e.g., an ACK. It is also used between CTAs because a DEV may be receiving or transmitting at the end of one allocation and be required to be in the opposite state at the beginning of the next CTA.

The backoff interframe space (BIFS) is only used in the CAP. It is based on the transmit/receive turnaround time and the length of time required to determine that the medium is clear. Thus, the BIFS is the sum of the SIFS and the time

> The BIFS acronym was not named for anyone involved in the 802.15.3 standards development. Rather, the acronym is intended to imply the name Biff, which, for some reason, the technical editor thought was funny.

required to perform a clear channel assessment (CCA).

The final IFS is called the retransmission interframe space (RIFS). A DEV that is expecting an ACK from another DEV will wait at least one REFS duration before it either resends the frame or gives up on it and sends a different frame. There were two viewpoints in the Task Group regarding the value of the RIFS. The first view was that the RIFS should be similar to the definition in 802.11; in which it is defined to be the amount of time to change from transmit to receive plus the time to listen for the ACK to start plus the time to change back to transmit from receive. Based on this viewpoint, the RIFS would be equal to SIFS plus CCA detect time plus SIFS, or equivalently, it would be equal to a BIFS plus a SIFS.

The other viewpoint was based on two assumptions. The first is that the there may be PHYs in the future that are unable to perform CCA or do it so slowly that it is unuseable. The other assumption is that it might be likely that the source DEV would miss the ACK starting and begin transmitting the next frame on top of the ACK being sent by the receiver. If this was true, then not only would the first frame transmission be lost, but also the second transmission would be lost as well. Using these assumptions, the RIFS would be equal to two SIFS plus the length time required for the ACK. If the DEV was requesting an Imm-ACK, then this time is easily calculated. On the other hand, if the DEV was requesting a Dly-ACK, then the source DEV cannot be sure of the exact duration of the

Dly-ACK frame. However, the source DEV can put an upper limit on this time and it would have to use the worst-case time for this method.

In the case of the 2.4 GHz PHY, the time required for CCA (7.3 µs) is much shorter than the time required for an Imm-ACK (about 23 µs), so the penalty for waiting for the entire ACK is quite large. In addition, the PHY preamble for the 2.4 GHz PHY has very good auto- and cross-correlation properties. These properties give the PHY preamble greater than 10 dB improvement of the signal-to-noise performance as compared with the data modulation. Thus, for the 2.4 GHz PHY, it is highly unlikely that a DEV would have reliable communications with another DEV in the piconet and not be able to detect the PHY preamble. Because there is a low probability of missing the preamble of the ACK and a large time penalty for waiting, the 2.4 GHz PHY uses the CCA method for determining RIFS. Because this IFS, like all of the others, is PHY dependent, the definition of it will be revisited when the next PHY, 802.15.3a, is selected.

Contention access period (CAP)

The CAP is used to provide efficient access for short bursts of data or commands. The PNC, however, is not required to allocate a usable CAP every superframe. If the PNC feels that the CAP traffic in the piconet is low enough, it could skip allocating the CAP to allow more CTAs in the superframe or to handle instances of superframe overloading (see "Managing bandwidth" on page 81). Unlike the channel time allocations (CTAs) in the CTAP, a DEV is not required to request time in the CAP before it is given a chance to use it. Because the DEV is not required to reserve time before using the CAP, it makes the CAP a very efficient method for transmitting small amounts of asynchronous data.

> All transmissions in the CAP either have to be ACKed in the CAP or sent as a no-ACK frame. The sending DEV needs to make sure that there is enough time remaining in the CAP for its frame and the anticipated ACK, if required. Dly-ACK is not allowed in the CAP, it can only be used during a CTA.

The PHY parameter, pBackoffSlot is the same value as the BIFS. In fact, BIFS could have been used instead of pBackoffSlot throughout the standard. This is a good candidate for things to fix in a Corrigendum.

When a DEV is going to transmit a frame in the CAP, it picks a random number between 0 and 7 to use as its backoff counter, n. It then waits until the channel has been clear for $n*$pBackoffSlot period of time and then gets to try and transmit its frame. Unlike 802.11, the counter is not reset if the medium is determined to be busy. Likewise, the counter is not reset for each superframe. One of the reasons for this choice is that the CAP is only a portion of the total time in the superframe. If the DEV picked a long backoff counter with a short CAP, it might never get a chance to transmit a long frame.

If the DEV transmits the frame but does not receive an acknowledgment, it may attempt to retransmit the frame. In this case, the backoff window increases from 0 to 7 to be 0 to 15. The window increases each time the frame is transmitted but not ACKed, next to 0 to 31 and finally to a window of 0 to 63. The DEV continues to use this backoff window until it gives up on sending the frame. Regardless if the previous frame was sent successfully or unsuccessfully, the next frame uses a backoff window of 0 to 7 for its first attempt to be transmitted during the CAP.

Channel time allocation period (CTAP)

The CTAP is defined as the time in the super-frame that follows the end of the CAP up until one SIFS before the start of the next beacon (which is one superframe duration after the start of the previous beacon). The PNC is not required to fill up the CTAP with CTAs or to expand the CAP to take up any unused time, so it is possible that there would not be any CTAs in the CTAP of a particular superframe. There might also be large gaps in time in between CTAs in the CTAP where the PNC has not allocated any channel time. The CTAs are communicated to the DEVs in the piconet via the CTA IE in the beacon. Each CTA IE contains one or more CTA blocks, each of

The CTAP was originally called the contention free period (CFP) throughout most of the life of the draft. However, the standard does allow the PNC to create open and association management CTAs (MCTAs) in the CTAP. Thus, the CTAP contains time when DEVs are contending for the medium using slotted aloha, and then so it could no longer be called a contention free period. This change was applied only in the latest drafts, so earlier documents in the task group will refer to this portion of the superframe as the CFP rather than the CTAP.

which has the source, destination, stream index, start time, and duration of the CTA. The CTA blocks are ordered by increasing start time of the allocation, with the lowest start time sent first over the air. Because IEs are limited to 255 octets in length, the PNC may be required to put multiple CTA IEs in the beacon to handle all of the allocations for the superframe. Only the source DEV is allowed to initiate a frame exchange sequence during a CTA. The destination is only allowed to transmit the requested ACK frames, either Imm-ACK or Dly-ACK, during the CTA.

> A CTA is the "property" of the source DEV. The source can use the channel time for almost any purpose, including making connections with the destination DEV using a non-802.15.3 protocol. This feature is used to support neighbor piconets.

Although every CTA has a stream index, the source DEV may send data with any stream index during the CTA. If the DEV has higher priority traffic it needs to send to the destination, it can use any or all of the CTAs that it has allocated to the destination to send the data. This includes sending asynchronous data or commands to the destination DEV. In general, once the DEV has been given the CTA, it is allowed to use the time for any purpose.

IEEE Std 802.15.3 always allowed, and in some cases required, the PNC to allocate channel time with another DEV in the piconet. In this case, the allocation is not for the PNC's DEV personality, i.e., using its DEVID, but rather for its PNC personality, i.e., using the PNCID. These CTAs are exactly the same as regular CTAs, except that the PNCID is either the SrcID or the DestID.[15] The standard refers to these types of CTAs as management CTAs (MCTAs). When the SrcID is either the PNCID or one of the DEVIDs of a DEV in the piconet, then the allocation is the same as other CTAs. However if the SrcID is either the unassociated ID (UnassocID) or the broadcast ID (BcstID), then the CTA is no longer contention free, but rather uses the slotted-aloha contention method for access. An MCTA with the UnassocID as the SrcID is called an association MCTA, whereas an MCTA with the BcstID as the SrcID is called an open MCTA. Open and association MCTAs were introduced to allow the possibility of a future PHY that was not able to do CCA in an efficient manner.

---

[15] These CTAs also use the MCTA Stream Index, but that does not affect the traffic in the CTA because frames with any valid stream index can be sent in the CTA as long as the SrcID matches the SrcID for the CTA.

The slotted-aloha mechanism is similar to the CSMA/CA method used in the CAP. Instead of using a backoff interval based on listening to the channel, the slotted-aloha method uses the entire MCTA as a backoff slot and counts down until it can access the medium. However, unlike the CSMA/CA method, the DEVs using slotted aloha do not have to listen first to see if the medium is clear, they just transmit when they have counted down to their slot. Consider two DEVs trying at the same time to access a single association MCTA that appears in every superframe. If it is the first time that the DEVs have attempted to send the Association Request command, they each choose an integer between one and two, inclusive (i.e., they pick either one or two). If the DEV has picked one, then it can attempt to transmit in this superframe. If the DEV has picked two, it decrements it by one for each association MCTA that is in the superframe and transmits when it reaches one.

> If there is more than one MCTA in a superframe and the DEV has multiple frames it needs to send to the PNC, the DEV could end up using more than one MCTA in the superframe. Once the DEV has sent a frame during an MCTA, it may begin the slotted-aloha backoff process as early as the next MCTA in that superframe. The backoff for the next frame starts with either 1 or 2 irrespective of the status of the backoff of the previous frame.

In the case of the two DEVs trying to associate, there is 50% chance that they will both choose the same number; thus, there is a 50% chance that they will collide in either the first or second superframe trying to send the Association Request command. If they do not collide, then both will be successful in sending the Association Request command within two superframes. If they collide in either the first or second superframes, then both DEVs will choose a new random number, this time between 1 and 4, inclusive, for their backoff. The range for the random number doubles with each failure until it reaches 256. The same process is used for open MCTAs where the DEVs are trying to send command frames, other than Association Request commands, to the PNC.

Another special type of CTA is the private CTA. This CTA is one where the source and destination DEVIDs are the same. This type of allocation is used to set aside time in the piconet for other types of communications, e.g., to allow time for dependent piconets or for the parent piconet. The dependent piconet is given a private CTA with its DEVID as the source and destination in the parent piconet's superframe. In the dependent piconet's superframe, the PNC provides a private

CTA with the PNCID as the source and destination in its own CTAP to set aside time for the parent piconet's operations.

Because the PNC is allowed to change the length of the CAP and location of the CTAs in every superframe, every DEV is required to hear the beacon before it transmits in the superframe. Thus, errors in the beacon reception will result in reduced throughput during the allocated CTA. To prevent the beacon FER from limiting the throughput of a stream, the standard allows a DEV to request a CTA, called a pseudo-static CTA, where the location of the allocation in the superframe is relatively constant between superframes. Because the PNC will not be moving the CTA, the DEV may continue to use this allocation for up to mMaxLostBeacons superframes, even if the DEV did not receive the beacon. The constant mMaxLostBeacons is set to a value of four in the standard. Thus, if the source DEV of a pseudo-static allocation only hears every other beacon from the PNC, it will not experience any reduction in throughput due to the lost beacons. If the source DEV had a regular CTA, it would have at least a 50% reduction in the throughput for the CTA.

The PNC is allowed to change the location of a pseudo-static CTA, but it needs to take care to ensure that enough time is allowed for the source and destinations of the pseudo-static CTAs to see the change in the allocation before the time is given to another DEV. Early versions of the draft used directed frames, first to the destination and then to the source, to communicate the change in the allocation. The PNC was not allowed to change the allocation until it had received ACKs to the directed frames from both DEVs participating in the CTA. However, because this method could potentially take a long time, especially if the PNC was having trouble communicating with either DEV, the task group decided to just use the beacon to communicate the change. The current process is very simple: when the PNC wants to move the pseudo-static CTA, it simply updates the allocation in the beacon to the new time. The PNC will not allocate the old time for the CTA to any other DEV until at least mMaxLostBeacons has passed. When the source DEV sees the new allocation in the beacon, it begins using the new allocation for its transmissions. When the destination DEV sees the new allocation, it begins listening during the new time and should also listen during the old time as well, in case the source DEV has not yet seen the new allocation. The destination DEV will stop listening during the old allocation and will begin receiving only during the new allocation after one of the following occurs:

a) the destination DEV receives a frame from the source DEV in the new allocation;

b) mMaxLostBeacons superframes have passed; or

c) the PNC has allocated the channel time to another set of DEVs.

If the destination sees the change in the allocation either before or at the same time that the source sees the change, no frames will be lost due to the change in the allocation. However, if the source sees the change before the destination sees the change, the source will be sending frames during a time when the destination is not listening, and the frames will be lost, requiring retransmission if the source is requiring ACKs. The destination DEV will either hear the new allocation within mMaxLostBeacons or it will have to begin scanning the channel to find PNC again. If the implementer wants the destination DEV to have the maximum probability of correctly receiving frames intended for it, then any time the DEV fails to hear the beacon, it should listen to the entire superframe for frames addressed to it. In this case, the DEV is allowed to ACK the frames that it receives, even if it has not heard the beacon because it the responsibility of the sending DEV to ensure that there is enough time left for the destination to send the ACK.

Comparing the contention access methods

IEEE Std 802.15.3 has defined two contention access methods. The original reasoning in adding slotted aloha as an access method was that UWB PHYs would be unable to perform clear channel assessment (CCA) in an efficient manner. However, the work of the TG3a alternative PHY task group has indicated that CCA for UWB PHYs could be done in approximately the same time as for the 2.4 GHz PHY. The requirements for this are the same as for the 2.4 GHz PHY, i.e., the CCA is based on either detecting the beginning of the PHY preamble at sensitivity or detecting any large energy in the receiver bandwidth. Under these assumptions, CCA for a UWB PHY could be done fairly efficiently.

Another argument for keeping open association MCTAs in the standard was the claim that slotted aloha could be as efficient or more efficient than a short CAP.[16] Depending on the traffic assumptions and the size of the CAP, it can be shown that associations MCTAs in some cases can offer less collisions and shorter average time to gain successful access to the channel. However, this analysis is based on the PNC allocating the minimum time that could be used to allow DEVs to associate in the piconet. The minimum usable CAP for association is one that allows time for just one Association Request command, the Imm-ACK, and two SIFSs. For the 2.4 GHz PHY at the 22 Mb/s base rate, the time required for the Association Request command, the Imm-ACK, and two SIFS is approximately 75 μs. If the PNC uses a CAP of this size, then the latency for an association can be longer than using a single association MCTA (which would be the same size) every superframe. However, if the PNC expands the CAP by even four backoff slots (e.g., 4 * 7.3 μs = 29.2 μs) to a total of 104 μs, then the CAP is more efficient on average than even two MCTAs, which would take up 150 μs.

For both contention methods, the PNC needs to determine the amount of time to be set aside for the DEVs to contend for the medium. In the case of the CAP, the PNC only needs to determine one number, the length of time to allocate for the CAP. One method to do this would be for the PNC to keep track of the ratio of the number of corrupted frames to the number of correctly received frames, regardless of the destination. Corrupted frames could be indicated by failed header check sequences (HCSs), frame check sequences (FCSs), or high receiver signal strength indication (RSSI) without a corresponding frame. Another method the PNC can use is to monitor the CAP to determine the length of time during which it was clear of transmissions. Regardless of the method used, if the PNC determines that the traffic during the CAP is high, it should increase the duration of the CAP. Conversely, if there is little traffic in the CAP and almost all of the frames are getting through, the PNC could reduce the duration of the CAP.

The task of determining the correct mixture of MCTAs is more complex for the PNC than is the task of determining the duration of the CAP. Not only is the PNC required to determine the number and frequency of MCTAs to allow for association, but it also has to determine the number and frequency of MCTAs to allow for other commands. The PNC must keep track of separate allocations for association and open MCTAs because Association Request commands may only be sent in an association MCTA and not in an open MCTA. Without the CAP, the PNC either needs to create uplink allocations to each of the DEVs in the piconet

---

[16] The slotted-aloha protocol has been shown to have a maximum theoretical throughput of only 36.8%. On the other hand, CSMA/CA protocols, such as the one used in IEEE 802.11 or in the CAP in IEEE 802.15.3, have been shown to have throughput in excess of 70%.

periodically to allow them to send commands to the PNC or it needs to allocate open MCTAs often for the same purpose. To do this, the PNC is required to keep track of the needs of the devices individually. If only a few devices are sending commands on a regular basis, then it would create uplink allocations to those DEVs while using an occasional open MCTA to allow the other DEVs a chance to communicate with the PNC.

Another issue with the use MCTAs to replace the CAP is that the allocation of MCTAs is a problem similar in complexity to managing sub-rate allocations. In the case of a sub-rate allocation, the requirements for the length and frequency of the allocation is known by the DEV requesting the channel time whereas for MCTAs, the PNC does not know the requirements of the DEV when it makes the allocation. Sub-rate allocations require the PNC to keep track of the frequency of the

> A sub-rate allocation is a channel time allocation that occurs less often than every superframe. For example, the allocation may occur every other superframe (CTA Rate Factor = 2) or every 8th superframe (CTA Rate Factor = 8). A super-rate allocation, on the other hand, occurs one or more times in every superframe. For example, a super-rate allocation that occurs twice in every superframe has CTA Rate Factor of 2. The Channel Time Request command differentiates between a sub-rate and a super-rate request with the CTA Rate Type field.

allocation and to make sure that the allocations are spread out among the superframes so that one superframe does not end up with more allocations than will fit in the allowed time. Once the PNC has determined the number of MCTAs to create and the timing for their allocation, it needs to find space in the superframe for the time allocation. In the absence of the CAP, all of the DEVs must explicitly request time to send asynchronous frames over the air. This increases the load on the PNC because it will have more CTRqs to handle as opposed to the CAP where the DEVs are directly responsible for getting time to send the data.

On the other hand, the slotted-aloha mechanism is easier, in some ways, to implement than the random access method in the CAP. Slotted aloha does not require that the DEV sense the medium before it transmits, which simplifies the transmit state machines. Accessing the medium during an open MCTA is similar to using an assigned CTA, when the DEV determines that it has access, it simply begins transmitting. Using the CAP requires that the MAC and PHY implementation are capable of efficiently implementing CCA, which may not always be the case for future PHYs. In addition, the slotted-aloha method also enforces a slot time that is understood by all of the DEVs in the piconet. When the CAP is used, however, it is not unusual for the DEVs in the piconet to calculate different slot boundaries because they do not all have the same probability of hearing a frame. When the timing for the slot boundaries among the DEVs is different, it may lead to more collisions and retransmissions in the CAP.

Guard time

In between every CTA is at least one SIFS duration. However, because the clocks of the individual DEVs are only synchronized once per superframe and they may have differing accuracies, the PNC is also required to set aside guard time to ensure that transmissions from one CTA do not overlap in time with transmissions in other CTAs. The worst-case scenario is when the clock of the source DEV in CTA n is slow and the clock of the source DEV in CTA $n + 1$ is fast. This is illustrated in Figure 4–4 where the ideal CTA positions are based on the DEVs having the same clock accuracy.

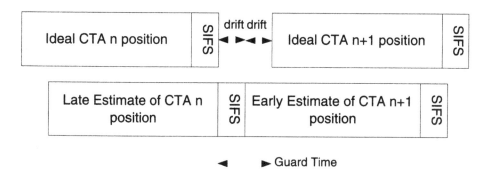

Figure 4–4: Illustration of worst-case CTA timing drift

The maximum drift that can occur is used to determine the guard time between successive CTAs in the CTAP and the time between the end of the last CTA and the start of beacon transmission. The drift is a function of the clock accuracy of the DEV and amount of time that has elapsed since the clocks were last synchronized. In 802.15.3, the

> A pseudo-static allocation can be moved by the PNC, but it needs to follow a special procedure (see "Channel time allocation period (CTAP)" on page 56). This procedure ensures that the DEV will still be able to use the old allocation for the maximum number of lost beacons that are allowed for this allocation.

synchronizing event is the beacon. For regular CTAs, the longest period of time between synchronizing events is the superframe duration. If the DEV does not hear the beacon, it may not transmit during the CTA and so it will not collide with other CTAs in the CTAP. In the case of pseudo-static CTAs, the DEVs may miss up to mMaxLostBeacons while still being allowed to transmit in the CTA. In this case, the maximum time between synchronizing events is equal to the product of mMaxLostBeacons and the superframe duration. The maximum drift for a single clock can be calculated with:

$$MaxDrift(interval) = \text{clock accuracy (ppm)} * 1,000,000 * interval$$

The clock accuracy for a DEV is PHY dependent and is ±25 ppm for the 2.4 GHz PHY. The interval is defined to be the time elapsed from the last timing event, i.e., the beacon.

With a maximum superframe duration of approximately 65 ms, the maximum drift between clocks for two DEVs in the piconet using regular CTAs is $(65 \text{ ms})*2*(25*10^{-6}) = 3.3 \mu s$. A summary of the maximum drift for regular and pseudo-static CTAs for various superframe durations is given in Table 4–1.

For typical superframe durations of less than 40 ms, the total drift for regular CTAs is pretty small, less than 2 μs, and so the PNC could add 2 μs to the 10 μs SIFS in between every CTA. However, if there are pseudo-static slots in the superframe, then choosing only a single guard time would almost double the amount of dead time on the air from 10 μs to 18 μs. In this case, it would be beneficial for the PNC to calculate the guard for each of the CTAs. Alternately, the PNC could calculate a finite number of guard times and then assign the appropriate guard time during the correct portion of the superframe. For example,

Table 4–1: Maximum clock drift for guard time calculation

Superframe duration	Max drift for regular CTA	Max drift for pseudo-static CTA
10 ms	0.5 μs	2.5 μs
20 ms	1 μs	5 μs
30 ms	1.5 μs	7.5 μs
40 ms	2 μs	10 μs
50 ms	2.5 μs	12.5 μs
65.535 ms	3.3 μs	16.4 μs

if the PNC was only going to use two guard times, it could assign CTAs in the first half of the super frame with one-half of the maximum drift added as guard time while CTAs in the second half of the superframe would have the maximum drift added as guard time. The GuardTime may be calculated as a function of the CTA intervals using:

$$\text{GuardTime(interval)} = [\text{MaxLostBeacons}(CTA_n) + \text{MaxLostBeacons}(CTA_{n+1}) + 2] * \text{MaxDrift(interval)}$$

The value of MaxLostBeacons for each CTA depends on whether the CTA is pseudo-static or dynamic. MaxLostBeacons is zero for a regular CTA and is equal to mMaxLostBeacons if the CTA is pseudo-static. The PNC can implement any method for allocating the guard time in the CTAP as long as it ensures that, even with the worst-case timing drifts, adjacent CTAs will always be separated by a SIFS.

> The CTA blocks in the beacon do not indicate that a CTA is either dynamic or pseudo-static. The DEV that requested the CTA and the PNC both know that it is a pseudo-static allocation because this is contained in the Channel Time Request command. The destination of the allocation will be informed of the type of allocation via the CTA Status IE that is placed in the beacon when the allocation is first granted. If the DEV misses this announcement, it can request the CTA Status IE from the PNC using the Probe Request command.

THE ROLE OF THE PNC

The PNC is key to the operation of the piconet. The DEV that takes on the role of the PNC is responsible for starting and stopping the piconet, admitting DEVs into the piconet, allocating channel time, changing the piconet channel if conditions deteriorate, and keeping the piconet parameters current. If the DEV supports security (Chapter 6), then it is also takes on the responsibility as the key originator for the piconet.

Starting a piconet

The first step in starting a piconet is to scan the available PHY channels to determine if there are other piconets in range. The DEV is also required to rate the channels based on interference. If there are no other piconets or if the DEV does not want to join a piconet, it may decide to start its own piconet. If the DEV is PNC capable, it will choose the channel with the least interference to start its piconet. The DEV will then listen to the selected channel to ensure that it is still quiet, i.e., that another piconet has not started on the channel or switched into the channel. When that scan is complete, the DEV will begin sending its beacon. At this point, the PNC will assign itself two DEVIDs: the PNCID for PNC related traffic and one of the unreserved DEVIDs for traffic that is intended for its application layers.

Once the PNC has started sending a beacon, the piconet has been formed, even though there is only one member. At this point, other DEVs may find the beacon and decide to join the piconet. The PNC will continue sending the beacon until it either hands over responsibility for the piconet to another DEV or it shuts down the piconet.

Handing over control

As DEVs join the piconet using the association process (see "Association" on page 74), they pass their capabilities field to the PNC. One of the purposes of this field is to tell the current PNC about the PNC capabilities of the new DEV. For example, the capabilities field indicates if the DEV is capable of becoming the PNC, if it is battery-powered, the number of associated DEVs it would support as PNC, and so on. The PNC looks at these fields and then determines if it is either required to or should hand over the control of the piconet to the new DEV. The order that these fields are compared in is given in Table 4–2.

As shown in Table 4–2, the PNC Des-Mode is the highest priority. In fact, it is the only entry in the table that requires the PNC to hand over its responsibility. The idea behind the PNC Des-Mode field is that there may be one DEV that the owner wants to be the PNC. In the case of the home, the person may want the DEV that is in the center of the house to be the PNC because that would maximize the range for the piconet. Users may select a DEV that they view as more capable, even if it is not judged to be so from the table above. The manufacturers may also set the PNC Des-Mode bit at the time of manufacture, e.g., in the receiver of a home theater system, due to the anticipated usage of the product. If the current PNC has the PNC Des-Mode field set to 1, then it does not have to hand over to a new DEV that does have the field set to 1. Otherwise, the PNC is forced to hand over responsibilities.

Table 4–2: Comparison order of fields for PNC handover

Order	Information	Note
1	PNC Des-Mode	If set to 1, the DEV prefers to be the PNC.
2	SEC	The DEV supports security as PNC is preferred.
3	PSRC	AC power as the DEV's power source is preferred.
4	Max associated DEVs	Higher number supported as PNC is preferred.
5	Max number CTRqBs	Higher number supported as PNC is preferred.
6	Transmitter power level	Higher transmitter power is preferred.
7	MAX PHY rate	Higher supported data rates is preferred.
8	DEV address	Tie breaker, no two DEVs have the same address.

If security is enabled in the piconet, then the PNC may not want to hand over to an new DEV if doing so would violate its security policies. In addition, if the PNC Des-Mode bit is not set, the PNC is not required to hand over to another DEV that rates higher based on Table 4–2. The standard only states that the DEV may hand over.

In addition to handing over to a new, more capable DEV that has joined the piconet, the PNC may wish to hand over control of the piconet if it is turning off. In either case, the PNC informs the DEV that it has been selected to be the new PNC by sending the PNC Handover Request command to that DEV. The complete frame exchange is illustrated in Figure 4–5. The PNC Handover Request command includes the number of DEVs currently in the piconet, the number of channel time request blocks (CTRqBs)[17] that it will send to the new PNC, and the number of PS Sets. The selected DEV is only allowed to refuse the hand-over request in two cases:

> A DEV is allowed to be the PNC of two piconets simultaneosly. However, this is significantly more demanding than running a single piconet, and so the DEV may wish to refuse a hand-over if it is already the PNC of a dependent piconet. If the DEV refuses the hand-over, it is possible that the current PNC will shut down the piconet because it is turning off and cannot find another DEV to take over as PNC. In this case, the DEV that is a dependent PNC may want to take on the added complexity of dual-PNC implementation to keep the dependent piconet operational.

a) The DEV selected to be the new PNC is currently the PNC of a dependent piconet and it is unable or unwilling to simultaneously run both piconets.

b) The DEV and PNC are in a dependent piconet and the selected DEV is unable to join the parent piconet either as a DEV or as a neighbor PNC.

Otherwise, the DEV is required to accept the nomination to become the new PNC and so it will then prepare to become the new PNC of the piconet.

The current PNC needs to give the new PNC all of the information about the associated DEVs in the piconet, the channel time requests, the power mode information, and any relevant security information. To accomplish this, the first command sent by the PNC is the PNC Information command. This command is a list of all the DEVs that are associated in the piconet and includes information about each of the DEVs and an indication if the DEV is a member of the piconet or if has only associated. A DEV is a member of the piconet if it is associated with the piconet and, if required, has established a secure relationship with the PNC.

---

[17] CTRqBs are the individual channel time requests that are sent by DEVs to PNC to secure channel time (see "Asynchronous data" on page 87 and "Stream connections" on page 92).

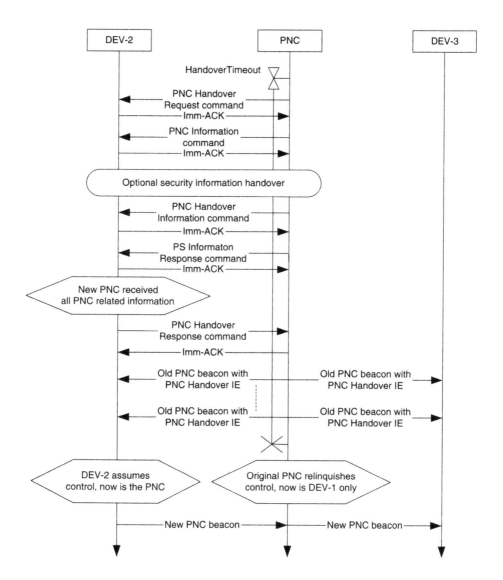

Figure 4–5: Frame exchange for PNC hand over

At this point, the new PNC may request the security information from the current PNC using the Security Information Request command. The Security Information commands that are sent in response to a Security Information Request command are intended to be used to handover authentication-related information about the DEVs that may or may not be currently members of the piconet. In particular, the Security Information command might contain a list of MAC addresses and associated information, e.g., cryptographic certificates, hashed keys, etc., of DEVs that should be allowed to join the piconet.

As an example, consider a DEV that has authenticated with the PNC but has left the piconet. The owner of the piconet and the DEV will have had to indicate to the PNC in some manner that the DEV should be allowed access to the piconet. The owner should not be required to re-authorize the DEV in the case of a PNC hand over. Instead this authorization should be passed in a trusted manner to the new PNC so that the DEV will automatically be re-authorized to become a member of the piconet as soon as it rejoins the piconet. The new PNC also needs this information because it will be required to re-authenticate with all of the DEVs in the piconet so that it can establish new management keys. Again, because some of this information normally requires user intervention in the authentication process, securely passing this information during the hand over process allows the hand-over to happen seamlessly. The standard does not define the content or format of this information because it depends on the authentication method that was used to allow DEVs to become members of the piconet. Authentication methods are outside of the scope of a MAC and PHY standard and so they are not specified in the 802.15.3 standard. A more complete description of the use of the Security Information command is given in "Security information" on page 199.

The next command that is sent is the PNC Handover Information command that contains all of the isochronous CTRqBs that the current PNC has received. Asynchronous requests are not passed, because some of them may expire before the hand-over is complete. The new PNC will use these requests to allocate channel time once it has taken over as PNC. The new PNC is also allowed to use the actual CTAs from the previous beacons to determine the CTAs for the beacons it will send once it has taken over. The last command that the old PNC sends to the new PNC is the PS Set Information Response command. This command contains the status of all of the DEVs in PS mode as well as all of the information about the PS sets (see "DSPS mode" on page 118). If there are no DEVs in PS mode, then the PNC does not send the PS Set Information Response command. Likewise, if there are no current isochronous CTRqBs, then the current PNC will not send the PNC Handover Information command. The new PNC knows that the current PNC will not be sending these commands because the number of CTRqBs or PS Sets in the PNC Handover Request command will be zero. Of course, it is not possible for there to be zero DEVs associated in the piconet. In the case of PNC hand over, there must be at least two DEVs, the current PNC and the DEV that will become the new PNC. Thus, the PNC Information command will always be sent in a PNC hand over.

Once the new PNC has received all of the information from the current PNC and it is ready to take over as the PNC of the new piconet, it sends the PNC Handover Response command. After the current PNC ACKs this command, it puts the PNC Handover IE in the beacon with the beacon number of the first beacon that will be sent by the new PNC. The last beacon that will be sent by the old PNC is the one that has a beacon number one less than the value of the Handover Beacon Number field. Once the PNC has ACKed the PNC Handover Response command, it must stop sending beacons at the time indicated for the new PNC to start sending its beacons. If for some reason the new PNC does not send beacons at the correct time, the old PNC is not allowed to begin sending beacons again. The reason for this is that it is possible that the old PNC is, for some reason, unable to hear the beacons being sent by the new PNC. If it begins to send beacons as well, it could either collide with the new PNC or confuse DEVs that are members of the piconet.

The new PNC is allowed to shutdown the piconet at any point in the hand-over process up until it places the PNC Handover IE in the beacon. The PNC might do this if it has been give a fixed length of time to shut itself down and the hand-over process is taking too long. In that case, it would change the process and simply shutdown the piconet.

Ideally, the hand-over process should not affect any of the data connections in the piconet. The new PNC is required to begin sending its beacon at the same time as when the old PNC was to have sent its beacon. Because the CTRqBs have been transferred to the new PNC, it is also able to continue to allocate channel time for all of the existing streams. However, the new PNC may not be able to support the same number of DEVs as the old PNC was able to support. In this case, the new

PNC will disassociate DEVs from the piconet until it has reduced them to a number that it can handle. Likewise, if the new PNC cannot support all of the CTRqBs that were handed over by the old PNC, it will terminate the streams of all of the allocations that it is unable to serve.The new PNC decides which DEVs are disassociated and which streams are terminated because it is the one that has best information regarding the streams and DEVs that it wants to keep in the piconet. The old PNC would have to randomly disassociate DEVs or terminate streams because it could not possibly know which ones the new PNC would like to keep.

The PNC is allowed to hand over control of the piconet to a DEV that is currently operating as a dependent PNC. Because the standard does not specify a method to merge (or to split) piconets, the implication is that the DEV that is the dependent PNC would be running two distinct piconets. Simultaneously running two piconets could well exceed

 The standard does not specify the method that a new PNC will use if it needs to reduce the number of DEVs or CTRqBs that it is retaining once the control of the piconet is handed over. This is an implementation-specific feature that can be used by an implementer to differentiate the quality of their MAC implementation. A correct answer to this problem is non-trivial.

the capabilities of a PNC-capable DEV. Thus, a DEV that is a dependent PNC can refuse the PNC hand over. A DEV refuses the PNC hand-over process by sending the PNC Handover Response command to the PNC with the Reason Code field set to "Handover refused, unable to act as PNC for more than one piconet."

Ending a piconet

If the PNC is shutting down and it does not have another PNC-capable DEV to which it can hand over responsibility or if it does not have time to complete a hand-over process, the PNC begins the shutdown procedure. The PNC cannot let all of the dependent piconets continue operation. If there are two dependent piconets in the current piconet, they would both desire to access to the rest of the channel time and yet they cannot both operate at the same time without the coordination of the old PNC. Because the PNC cannot hope to make an intelligent decision about which dependent piconet to leave control, it simply picks the dependent piconet with the lowest DEVID to remain operating in the channel. Because neighbor DEVIDs are larger than all of the assigned DEVIDs, the neighbor PNC would be selected only if there were not any child PNCs in the piconet.

The PNC informs the chosen dependent PNC as well as other DEVs in the piconet via the PNC Shutdown IE in the beacon. This IE contains the DEVID of the dependent PNC that will be allowed to continue operation in the channel. All of the other dependent PNCs are required to start their own shutdown process as well. Once the parent PNC has completed its   shutdown procedure, the selected dependent PNC is allowed to use all of the time in the channel for its own piconet.

> There was some discussion in the Task Group as to the best method for determining which dependent piconet would "survive" the shutdown of the parent piconet (once it was determined that collisions could happen). The reality is that any decision is somewhat arbitrary and so the group consciously chose an arbitrary criteria, the DEVID of the dependent PNC. There really is no deep logic behind this decision except that it does have the nice result that it automatically puts neighbor PNCs at the end of the list, which is likely a desirable outcome.

If the PNC was shutting down because it did not have time to hand over and there are no dependent piconets, one of the remaining PNC-capable DEVs in the piconet is allowed to start a piconet in the same channel using the piconet start procedure. However, if there is a remaining dependent piconet, then that DEV would have to either find another channel to start a new piconet or join the remaining piconet and try to start its own dependent piconet. Other DEVs may also try to join the remaining piconet, or they may search the other PHY channels to find another piconet to join.

JOINING AND LEAVING THE PICONET

The PNC controls membership in the piconet using association, disassociation, and secure membership. The method that the PNC uses to determine which DEVs are authorized to join the piconet with secure membership is outside of the scope of the standard. Authorization will normally involve the intervention of the user to verify that a new DEV will be allowed to join the piconet and potentially may involve cryptographic means to identify or authenticate the new DEV.

Association

The association process accomplishes two things. The first is to assign a DEVID to the DEV. Although every DEV has a unique DEV address, the 8-octet address is too long for the frame header. Thus, the PNC assigns a shorter, 1-octet DEVID to the new DEV in the association process. The second thing the association process accomplishes is to limit the requirements on the current PNC. For each DEV that joins the piconet, the PNC is required to store some information and to maintain a timer that keeps track of its activity. Because

> The DEV address used in 802.15.3 is based on the 8-octet MAC address assigned by the IEEE registration authority committee (RAC). Up until late in the sponsor ballot process, the 802.15.3 draft used the older 6-octet MAC address as the DEV address (802.11, for example, uses this type of MAC address). Because 802.15.3 was a new standard and it did not require MAC level routing of frames, the Task Group chose to use the new, longer MAC address offered by the IEEE. Companies and organizations can request groups of addresses from the IEEE RAC, who ensures that these numbers are unique among all network adapters.

the resources of the PNC are limited, it is important that the PNC is able to refuse to admit new DEVs once it has reached its limit. The PNC may also uses the association process to refuse to allow neighbor piconets in the current piconet. A DEV that wishes to join as a neighbor must state its intention to do so when it associates. The PNC may decide at this point to refuse to allow the DEV to join as a neighbor.

Because a DEV that wishes to associate has not been assigned a DEVID, it uses the unassociated ID (UnassocID) for all of its communications until it has been assigned a DEVID by the PNC. All of the associating DEVs share a single DEVID, and so they must use a contention-based access method to send the Association Request command to the PNC. In the 2.4 GHz PHY, the CAP is always available for sending Association Request commands. In addition, the PNC may allocate association MCTAs in the CTAP to allow DEVs to send Association Request commands to the PNC. The complete frame exchange sequence for association is illustrated in Figure 4–6.

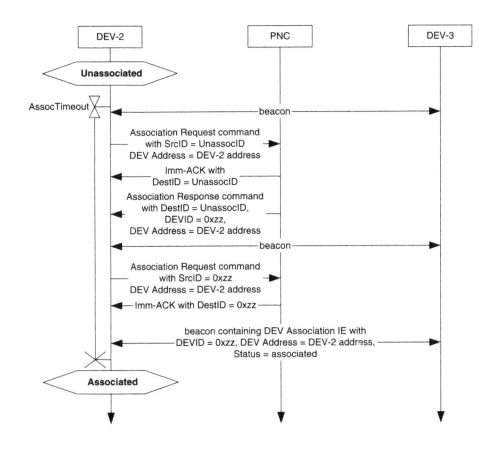

Figure 4–6: Frame exchange a DEV associating with a PNC

The Association Request command contains the DEV address of the new DEV as well as its Overall Capabilities field, and its requested association timeout period (ATP). If the PNC receives this command successfully, it sends an Imm-ACK with the UnassocID as the DestID. The PNC then responds to the Association Request command with an Association Response command that either accepts or rejects the request for association. Because the DEV has not yet learned its new DEVID, the Association Response command is sent with the UnassocID as the DestID and the ACK policy is set to no-ACK. This prevents multiple DEVs from attempting to ACK the frame at the same time. If the PNC accepts the association, it sets the DEVID to the newly assigned ID, responds with the ATP that it is willing to accept, and includes the DEV address of the DEV to which it is responding. The DEVs that receive this command look for their DEV address to determine if the PNC is responding to their request. Additionally, the PNC is able to include a Vendor Specific IE in the response. This IE might contain information regarding the authentication method that is being used by the PNC for secure membership.

To acknowledge that it received this frame, the DEV resends the Association Request command, this time with its new DEVID as the SrcID for the frame. If the PNC successfully receives this frame, it sends an Imm-ACK to the newly associated DEV and places the DEV Association IE in the beacon. This IE informs the other DEVs in the piconet that a new DEV has joined the piconet. However, if security is enabled in the piconet, the other DEVs need to wait until the new DEV completes the secure membership process before they can begin communications with the new DEV. The DEVs in the piconet will be informed that the new DEV has become a member of the piconet when the PNC broadcasts the PNC Information command with the new DEVID as an entry and the Membership Status bit set to indicate secure membership.

If the PNC is rejecting the request, it responds with an Association Response command with the DEVID field set to the UnassocID and the DEV address field set to the address of the DEV for which it is rejecting association. The PNC also includes a Reason Code that explains why the PNC is not going to allow the DEV to associate. Reasons why the PNC may refuse association are as follows:

• The PNC is already serving its maximum number of DEVs.

• There is not enough channel time available to serve the DEV.

• There is too much interference in the channel.

• The PNC is turning off with no PNC-capable DEV in the piconet to hand over responsibilities.

• The PNC will not allow the formation of neighbor piconets.[18]

• The PNC is in the process of changing the PHY channel.

---

[18] This reason is only valid if the DEV has indicated in its capability field that it is associating as a neighbor PNC.

- The PNC is in the middle of a hand-over process.
- The PNC is denying association for some unspecified reason.

The last reason is a catch-all because the PNC can reject an association request for just about any reason. Depending on the reason why the association was refused, the DEV may want to try again to reassociate. For example, in the case of channel change or hand-over, the DEV can watch the beacon to see when the process has finished and attempt to associate at that time.

Broadcasting piconet information

The PNC broadcasts the piconet information using the PNC Information command after a DEV becomes a member of the piconet. This means that if secure membership is required for the piconet, the PNC will wait until after the secure membership process is complete to broadcast the piconet information with the new DEV. In addition, the PNC is required to send the piconet information for each of the DEVs that are members of the piconet at least once every mBroadcast-DEVInfoDuration[19] via a PNC Information command. The purpose of broadcasting the PNC Information command is threefold:

a) The first purpose is to inform all of the members of the piconet that a new DEV has joined. Note that the DEV Association IE is sent when the DEV finishes the association process and it only has a portion of the DEV's capability field. In addition, if security is required for the piconet, the other DEVs need to wait until the new DEV completes the secure membership process with the PNC before they can communicate with the new DEV. The PNC Information command will only contain the new DEVs information once it has successfully become a member of the piconet (i.e., it has been given a management key for the PNC; see "Security policies" on page 193).

b) The second purpose of the broadcast PNC Information command is that it provides the DEVIDs, DEV Addresses and capabilities fields for the other members of the piconet. Thus, once a DEV joins an 802.15.3 piconet, it will immediately find out the information required to begin communications with other DEVs in the piconet.

---

[19] mBroadcastDEVInfoDuration is a constant defined at the end of Clause 8 in the standard with a value of 64*mMaxSuperframeDuration or just a little under 4.2 s.

c) The final purpose of the broadcast PNC Information command is that it provides a method by which DEVs in the piconet can update their internal list of the members of the piconet. For example, if a DEV is in a power-save mode, it might miss the association and disassociation announcements that occur in the beacon. Once the DEV receives the broadcast PNC Information command, it can quickly update its internal memory with information about the current members of the piconet.

Disassociation

A DEV ends its relationship with the PNC and the piconet via the disassociation process. The frame exchange sequence for a DEV initiating the disassociation process is illustrated in Figure 4–7, whereas the process of the PNC disassociating a DEV is illustrated in Figure 4–8. If the DEV wishes to disassociate, it sends the Disassociation Request command to the PNC. When the PNC acknowledges the receipt of this command, the DEV considers itself disassociated. Likewise, if the PNC wants to disassociate a DEV, it sends the Disassociation Request command with the appropriate Reason Code to the DEV. When the DEV ACKs this command, the PNC will consider the DEV to be disassociated.

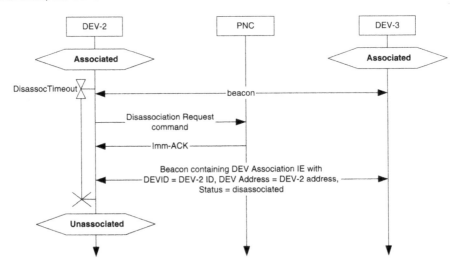

Figure 4–7:   Frame exchange for a DEV dissociating

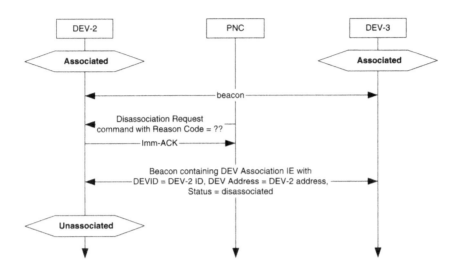

Figure 4–8:   Frame exchange for the PNC dissociating a DEV

Valid reason codes for the Disassociation Request command are as follows:

• The DEV's ATP has expired.

• The PNC is having problems with the channel and is unable to continue serving the DEV.

• The PNC is unable to service the DEV.

• The PNC is turning off with no PNC-capable DEV in the piconet.

• The DEV is leaving the piconet.

> The PNC may disassociate a DEV because it cannot support the number of DEVs that are currently associated with the PNC ("PNC unable to service DEV"). On the other hand, the PNC may be able to handle the DEV as a a member of the piconet but it would not be able to allocate any additional channel time to the DEV. In this case, the PNC would disassociate the DEV with the reason code "Channel too severe to service the DEV."

The first four reason codes are only valid when the command is sent by the PNC, whereas the last reason code is only valid when the DEV is sending the command. The Disassociation command is also used to stop a neighbor piconet. However, the PNC is required to perform some additional steps for this process, as described in "Parent PNC ceasing operations with dependent piconets" on page 174.

Because it is possible or even likely that DEVs will move out of range of the PNC without the PNC being aware of it, the standard requires that a timeout interval be kept for every DEV in the piconet. This interval, the ATP, is requested by the DEV during association. However, the PNC has the final say on the length of time allowed for the ATP. Because the ATP is a 2-octet value in milliseconds, the longest possible ATP is approximately 65 s. If the PNC has not received a frame from the DEV within the ATP duration, it considers the DEV disassociated from the piconet. In order to assure that the PNC has received a frame, the DEV should send at least one frame to the PNC within the ATP with the ACK policy set to be Imm-ACK. If a DEV does not have information that it needs to send to the PNC, it can send a Probe Request command that does not request any IEs to the PNC with the ACK Policy field set to Imm-ACK to renew its ATP timer.

Likewise, if the DEV has not heard from the PNC for an ATP duration, i.e., it has missed quite a few beacons, it considers itself disassociated from the piconet and stops using the DEVID that was assigned to it by the PNC. The DEV may look for another piconet to join, or if it is capable, it could start its own piconet as the PNC.

Assigning DEVIDs

The PNC is required to assign DEVIDs in increasing order, except when it wishes to reuse the DEVID of a DEV that has left the piconet. Thus, when a DEV starts a piconet, it takes on the DEVID 0x00[20] as the PNC and the DEVID 0x01 for its DEV personality. Because the DEVIDs are reused, the PNC must make sure that the DEV that previously had the DEVID has been away long enough that it knows that it has been disassociated from the piconet (see "Disassociation" on page 78). To be sure, the PNC needs to wait until at least two times the ATP duration for that DEV has passed.

The DEVIDs are assigned as follows:

- *0x00:* The PNC (PNCID)

- *0x01 to 0xEC:* IDs for regular members of the piconet (DEVIDs)

- *0xED to 0xF6:* Reserved IDs for future use

---

[20] This book and the 802.15.3 standard use the notation 0x to indicate that the numbers that follow are in hexadecimal notation.

- *0xF7 to 0xFC:* IDs for the PNCs of neighbor piconets (NbrID)

- *0xFD:* The multicast ID (McstID)

- *0xFE:* The ID for unassociated DEVs (UnassocID)

- *0xFF:* The broadcast ID (BcstID)

MANAGING BANDWIDTH

Once the piconet has been started and a DEV has joined, the next task is to open communications with the other DEVs in the piconet. The method that the DEV uses to do this depends on the type of data and the length of the CAP in the superframe. The DEV also needs to determine which ACK policy it will use in sending the data because this affects when the data can be sent and the amount of time that may need to be allocated for sending the data. If the DEV needs to request channel time for the data, it will choose either an isochronous or asynchronous request method.

Acknowledgements

IEEE Std 802.15.3 provides four acknowledgment (ACK) policies that can be used for data. The ACK policy is a field in the MAC header that tells the receiving DEV what type of ACK the source DEV is requesting. The four types of ACK policies are:

a) No acknowledgment (no-ACK)

b) Immediate acknowledgment (Imm-ACK)

c) Delayed acknowledgment (Dly-ACK)

> The ACK policy field is 2 bits and it was arranged so that each bit has a different meaning. The second bit in the field indicates if a response is required. If the second bit is set to zero, then the ACK policy is either no-ACK or Dly-ACK and the receiver will not respond to the frame. On the other hand, if the second bit is set to one, then the destination is supposed to respond, either with an Imm-ACK frame or Dly-ACK frame, if it correctly receives the frame. This allows the MAC implementation to check only a single bit to determine if it will need to switch from receive to transmit mode following the reception of the current frame.

d) Delayed acknowledgement request (Dly-ACK request)

The no-ACK policy is used when there are multiple DEVs that may wish to ACK the frame, e.g., for the Association Response command, for broadcast frames, and for multicast frames. In addition, the data may not require an ACK, especially if it is time-sensitive data like streaming video, audio, or two-way voice traffic. In that case, a retransmission may arrive too late and the upper layers will handle dropped frames. In other instances, the ACKs will be handled by a higher layer protocol. For example, the delayed ACK mechanism in this standard could also be implemented at a higher layer by using no-ACK for the data with either a short CTA in the reverse direction or even a sub-rate slot in which the destination DEV reports the frames that it has received.

The Imm-ACK policy is used when the DEV needs to confirm that the destination has received the frame and may want to attempt a retransmission if it fails. The Imm-ACK policy is used for most of the command frames that are sent to a single DEV because the protocol state machines rely on knowing if the message was delivered. The Imm-ACK policy is also appropriate for typical data transfers where even a single missing octet will render a file transfer useless. Because the Dly-ACK policy cannot be used during the CAP or during an asynchronous CTA, a DEV would need to use the Imm-ACK policy during these times if it needs to verify that the destination DEV correctly received the information. When a DEV receives a frame with the Imm-ACK policy set, it verifies that the calculations for the header check sequence (HCS) and frame check sequence (FCS) match the data in the frame. If both checks pass, the DEV sends the Imm-ACK frame one SIFS following the frame that requested the Imm-ACK. Note that it is the responsibility of the sending DEV to make sure that there is enough time remaining in either the CAP or CTA for the destination to send the Imm-ACK frame.

The term MSDU refers to data that is passed between the MAC and higher protocol layers. A MAC protocol data unit (MPDU), on the other hand, refers to data units that are passed between the MAC and the PHY. A more in-depth discussion of the differences between an MSDU and an MPDU is given in "Fragmentation/defragmentation" on page 96.

The Imm-ACK frame indicates that the destination received a frame for which it was the DestID and that the HCS and FCS were valid. The destination may still discard the frame depending on other factors. For example, if the Integrity Code check or a secure frame fails (see "Overview of AES CCM" on page 195), the DEV may discard the frame. The DEV may also discard the frame if its internal buffers are full. If the DEV receives only some of the fragments of a MAC service data unit (MSDU), it will ACK the fragments it receives, but it will discard them if it does not receive all of the fragments. Thus, the higher layer protocols need to be aware that the ACK alone will not guarantee that the entire frame was passed up from the MAC.

The third type of ACK policy is the Dly-ACK policy. This policy is used in conjunction with the Dly-ACK request policy to set up and operate an ACK method in which the destination DEV stores ACK information until it is requested by the source. The reason for including Dly-ACK in the standard is that the Imm-ACK method is relatively inefficient in sending data due to the fixed overhead of the two SIFS and Imm-ACK frame required to send what is really one

> The capability provided by Dly-ACK is sometimes referred to as a group ACK or burst ACK in other standards or specifications. The naming of the Dly-ACK functionality is arbitrary and could have been called Grp-ACK (group ACK) instead. However, the first versions of the draft used delayed ACK as the name and the task group decided to stick with this name for the final draft. The abbreviation did change; the delayed ACK policy was initially abbreviated as Del-ACK.

bit of information (i.e., "Did the frame get through?"). The inefficiencies are particularly pronounced for higher data rates, e.g., for data rates faster than 44 Mb/s.

The improvement in throughput from using Dly-ACK is shown in Table 4–3. The 110 Mb/s and 480 Mb/s data rates are not defined for the current PHY, but they are included in the table to provide a comparison for what might happen with the data rates proposed for 802.15.3a.

Table 4–3: Throughput improvement using Dly-ACK

PHY data rate	Rate with Imm-ACK	Rate with Dly-ACK	Improvement
11 Mb/s	10.4 Mb/s	10.7 Mb/s	2.4%
22 Mb/s	20.1 Mb/s	21.0 Mb/s	3.8%
33 Mb/s	29.1 Mb/s	30.7 Mb/s	5.6%
44 Mb/s	37.4 Mb/s	40.1 Mb/s	7.2%
55 Mb/s	45.0 Mb/s	48.9 Mb/s	8.8%
110 Mb/s	84.8 Mb/s	94.9 Mb/s	11.4%
480 Mb/s	209 Mb/s	280 Mb/s	33.9%

In the calculations for Table 4–3, the frame size that was used is pMax-FrameBodySize and the number of frames in a burst is 10. The PHY dependent parameter pMaxFrameBodySize parameter is the limit on the size of a frame that can be sent with a given PHY. For the 2.4 GHz PHY, this is set to 2048 octets (which includes the FCS, but not the PHY preamble, PHY header, MAC header, or HCS). Note that the rates in the table are only the

> Constants used in the standard are prefixed either with a lower-case "m" for MAC constants or a lowercase "p" for PHY constants. The MAC constants are defined in a table in subclause 8.15 of the standard. The PHY constants are also summarized in a table in this subclause, but the actual definitions are found in the 2.4 GHz PHY definition in clause 11. In PDF copies of the draft, each location of these constants in the text is a hyperlink to the definition so that clicking on the constant should cause the PDF display to jump to the actual definition.

instantaneous delivered data rates and do not include the overhead due to the beacon, CAP, guard times, or the presence of other DEVs in the piconet. The Dly-ACK mechanism is not particularly compelling for the 11 Mb/s data rate, but it becomes more interesting at the higher data rates. At the 55 Mb/s data rate, using Dly-ACK provides an extra 4 Mb/s of instantaneous data rate. The improvement is greater for shorter frames because a larger percentage of the time is spent sending the headers and waiting for the PHY to switch from transmit to receive and back again.

Because the Dly-ACK policy places a burden on the finite resources of the destination DEV, the source DEV is required to negotiate the parameters of the Dly-ACK burst before it can begin using the Dly-ACK method. A burst is the collection of the frames that are pending acknowledgment via a Dly-ACK frame. Because of this, the burst may be spread across multiple CTAs or even across multiple superframes. The Max Frames field in the Dly-ACK frame is the number of frames that the destination is willing to receive in a single burst, regardless of size. In addition, the destination needs to specify the maximum amount of data, in octets, that it can receive in a single burst. This number is communicated in the Max Burst field, which is the number of pMaxFrameBodySize frames that it can receive in a single burst.

The Max Frames field allows the destination to limit the number of "handles" or "bins" that it uses to store the incoming frames. The DEV may have a limit to the number of memory addresses that are assigned for incoming frames. The Max Burst field allows the destination to limit the impact of Dly-ACK based on the overall size of its buffer. These two parameters together give the destination control over the amount of data it needs to store to participate in the Dly-ACK mechanism.

The source DEV begins the Dly-ACK negotiation by sending a frame with ACK policy set to Dly-ACK request. If the destination does not want to support the Dly-ACK method, it responds with an Imm-ACK frame, as shown in Figure 4–9. If it is willing to use the Dly-ACK method, it responds with a Dly-ACK frame that contains the Max Burst and Max Frames that it is willing to support, as shown in Figure 4–10.

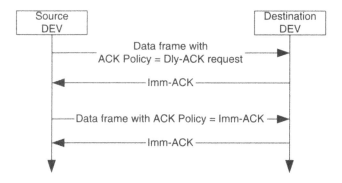

Figure 4–9:   Frame exchange for an unsuccessful Dly-ACK negotiation

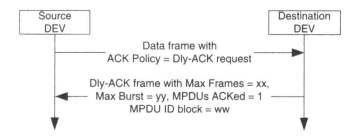

Figure 4–10:   Frame exchange for a successful Dly-ACK negotiation

If the destination accepts the request for the use of Dly-ACK, the source is then able to begin sending frames to the destination with the ACK policy set to Dly-ACK. The source DEV can send up to either the Max Burst or the number of

Max Frames to the destination. The source then requests the ACK information by sending a frame with the ACK Policy set to Dly-ACK request. The destination responds with the MAC service data unit (MSDU) number and fragment number of all of the frames that it successfully received. The Dly-ACK frame also contains a new Max Burst and Max Frames values that the source destination may use to continue the Dly-ACK process.

This process is illustrated in Figure 4–11. The source first requests the use of Dly-ACK and is given a positive response that allows up to two frames to be sent. In the second data frame that is sent, the source sets the ACK Policy to Dly-ACK request and receives a response from the destination that it received MSDU number 3 but did not receive MSDU number 2. In the figure, the MSDUs are not fragmented and only the MSDU number and not the fragment number is listed for clarity. Following the reception of the Dly-ACK frame, the source resends MSDU number 2 as well as MSDUs numbered 4 and 5 and requests the Dly-ACK frame with the frame that contains MSDU number 5. The destination responds with a Dly-ACK frame that acknowledges all three MSDUs.

If the destination is participating in a Dly-ACK process with one DEV and it receives a request from another DEV, it can determine from its internal resources if it is able to support the other DEV at that time. Because the Dly-ACK process is renegotiated every time a new burst is started, the destination is in control of the amount of resources it has promised to the source DEVs with which it is communicating.

If the destination replies with a Dly-ACK frame with the Max Burst field equal to zero, then the source needs to stop transmitting in the current CTA, but it is allowed to restart the Dly-ACK process in the next CTA by sending a frame with the ACK policy set to Dly-ACK request. If the source does not receive a Dly-ACK frame in response to its request, it will repeat the last frame that it sent until it gets an ACK for the frame. The source DEV may also send an empty data frame with ACK policy set to Dly-ACK request to get the receiver to send the Dly-ACK frame. One reason to do this is that an empty data frame has a higher probability of getting through than a long data frame.

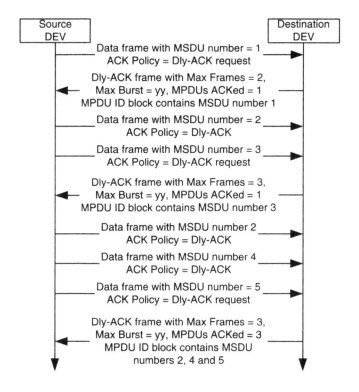

Figure 4–11: Data exchange using Dly-ACK

Asynchronous data

IEEE Std 802.15.3 differentiates between asynchronous and isochronous data connections. An asynchronous data connection is one in which it is OK if there are pauses in the transmission of the data or where a short latency is not required in sending the data. Examples of applications for asynchronous data connection include web surfing, file downloads, backing up data, downloading email, and so on. Although the standard explicitly refers to only one asynchronous allocation method, there are really a total of three ways in which asynchronous data may be communicated in the piconet:

a) Send the data during the CAP[21]

b) Request channel time for an individual asynchronous allocation

c) Request channel time for a group asynchronous allocation

If the amount of data to be transferred is very small or if the DEV needs to send a command to either the PNC or another DEV, it can use the CAP to send the frame. In the 2.4 GHz PHY, the PNC is required to allow the CAP to be used for data transfer. Using the CAP is a very efficient means for sending a small number of data frames or commands.

If the amount of information that the DEV has to send is large enough, if the CAP is too small, or if another PHY allows the PNC to forbid data transfer in the CAP, then the DEV should use the Channel Time Request (CTRq) command to get time to send the data. Because the same CTRq command is used for both asynchronous and isochronous requests, the PNC uses

> One way to understand the difference between an isochronous request and an asynchronous request is that an isochronous request is a request for a data rate, e.g., for a request for 5 Mb/s, whereas an asynchronous request is a request for a specific amount of data to be sent, i.e., a request to send 30 Mb of data.

the stream index to differentiate between the two requests. An asynchronous request sets the stream index field to the asynchronous stream index, 0x00. The allocation of asynchronous time is different than that of isochronous time. A isochronous CTRq is requesting that the PNC allocate time in the superframe continually until the source, destination, or PNC request its termination. An asynchronous CTRq, on the other hand, is a request for a total amount of channel time to be allocated when it is convenient for the PNC.

There are two types of asynchronous allocations that the DEV can request from the PNC. The first type is the individual allocation request where each CTRqB corresponds to a request for a CTA to a single DEVID. Using this method, a DEV that needs asynchronous allocations to four different destinations would include four CTRqBs in the CTRq command that it sends to the PNC, one for each of the destinations. The PNC will then allocate the channel time when it becomes

---

[21] If the PNC indicates in the beacon that data is allowed to be sent in the CAP. For the 2.4 GHz PHY, the PNC is required to allow DEVs to use the CAP for data.

available with a different CTA for each of the destinations specified in the CTRqBs.

The second type is called a group allocation request. It is a request for a single CTA with multiple destinations. Because the CTA IE does not support multiple destinations, other than the multicast and broadcast addresses, the PNC creates this allocation by placing CTA blocks in the beacon that have the same SrcID, stream index (= asynchronous stream index), start time, and duration, but different DestIDs. Because the SrcID is the same for all of the CTA blocks, there will not be contention for the channel during that CTA. However, all of the DEVs that are listed as destinations will have to listen to the entire CTA in case a frame directed to the DEV is sent.

A DEV is only allowed to use one of the two methods at a time. If the DEV sends a CTRq command with an asynchronous CTRqB where the allocation method is different from its previous requests, the PNC will drop all of the prior asynchronous requests from the DEV. It will then consider the current request using the method indicated in the CTRqB.

The frame exchange for negotiating an asynchronous CTA is illustrated in Figure 4–12. The DEV begins the request process by sending a CTRq command with an asynchronous CTRqB to the PNC. The PNC does not need to refuse an asynchronous request if it is unable to fulfill the request at this time. Instead, the PNC should retain the request and wait until there is time available in the superframe for the allocation. When the PNC has time available in a superframe for the allocation, the PNC will put the appropriate CTA blocks in the beacon for the asynchronous allocation. Unlike a typical isochronous allocation, the asynchronous request is generally prompted by data arriving at the MAC/LLC interface and not by a specific request from a higher layer to allocate time for communications. Thus, the asynchronous CTRqBs can be thought of as being the current status of the transmit queue for the source DEV.

Because the asynchronous allocation is a transient one, there is not a special process for modifying an existing asynchronous CTRq. Instead, each new request replaces the old one. In that sense, subsequent CTRq commands with asynchronous CTRqBs can be thought of as a modification to the existing allocations.

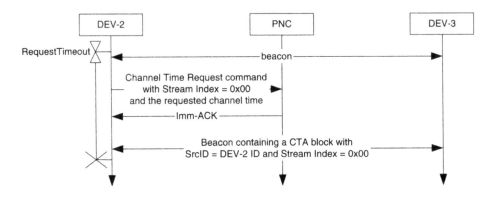

Figure 4–12: Frame exchange for asynchronous CTRq

It is possible for either the source or the PNC to terminate an asynchronous allocation. Because asynchronous allocations are transient, the destination is not allowed to request that they are terminated. If the source is sending very short bursts of asynchronous data, then it would be unlikely that

> The CTRq command originally had a Stream Termination bit. However, it was determined that this field was superfluous because the PNC will know that the DEV is requesting the termination of a stream when the DEV sends a CTRq that requests zero channel time. Because it does not make any sense to allocate zero time, this can only be a request to terminate the stream. This is true for both asynchronous and isochronous allocations.

either the source or the destination would be able to get the termination command to the PNC before the allocation expires on its own. If the source is sending a large amount of asynchronous data, e.g., a file transfer via FTP, then there may be time to terminate the allocation if the source feels it is necessary. For example, if the higher-layer application sending the data stopped in the middle of the transfer, the source DEV would want to terminate the asynchronous allocation to free up the bandwidth in the network. When the source DEV wants to terminate an asynchronous allocation, it sends a CTRq command to the PNC with an asynchronous CTRqB that has the Target ID List set to the list of destinations whose allocation is being cancelled and the Stream Index and all other fields are set to zero. When the PNC receives this request, it will ACK the command and stop allocating channel time for the asynchronous request. Note that if the original request used the group allocation method, then this termination applies to all of the destination DEVs. The frame exchange sequence for the source DEV terminating an asynchronous allocation is illustrated in Figure 4–13.

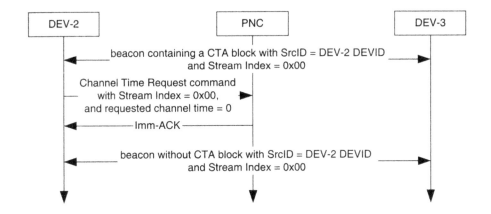

Figure 4–13:   Frame exchange for the source DEV terminating

The PNC may also wish to terminate an asynchronous allocation. The PNC's resources may have become overloaded, and so it is unable to handle the asynchronous allocation. If the destination DEV is disassociated from the piconet, either by its own request, by ATP timeout, or because the PNC requests it, the PNC may want to explicitly terminate the allocation and inform the source DEV. When the PNC wants to terminate an asynchronous allocation, it simply stops allocating the channel time and sends a channel time response command to the source DEV. The reason codes that the PNC might send when it terminates an asynchronous allocation include the following:

- Target DEV unassociated
- Target DEV not a member
- Stream terminated by PNC

The PNC would not normally terminate an asynchronous allocation because there is not currently enough channel time. Because it is an asynchronous allocation, it will hold on to the request until it gets a chance to fulfill it. One consequence of saving the pending requests is that the PNC may run out of space to store the CTRqBs. In this case, the PNC would need refuse any new CTRqs until it is able

to reduce the number of pending requests. The frame exchange for the PNC
terminating an asynchronous allocation is illustrated in Figure 4–14.

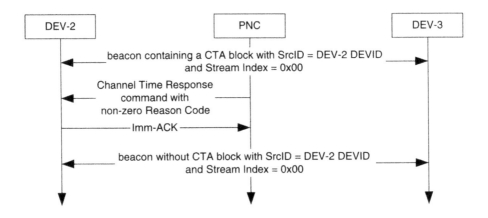

**Figure 4–14:   Frame exchange for the PNC terminating
an asynchronous channel time request**

Stream connections

One of the key features of the 802.15.3 protocol is the ability of a DEV to request
that the PNC allocate a specific length of time in each superframe or in every n^{th}
superframe for the DEV to send frames. Because these allocations occur at regular
intervals, they are called isochronous allocations. Isochronous applications need a
guaranteed data rate and latency to support the quality of the user experience. A
wireless connection, on the other hand, deals only with channel time, frame error
rates (FERs), and instantaneous bit rates. One of the key issues in creating
connections that support QoS is to map the requested bandwidth and latency into
channel time and its repetition, i.e., the latency between allocations.

IEEE Std 802.15.3 initially used a three-way negotiation among the source, the
PNC, and the destination. This allocation request included the maximum transmit
delay variation, minimum data rate, peak data rate, average data rate, maximum
burst size, average frame size, maximum retransmission duration, and receive
window size [B16]. The PNC would use these application level requirements to
determine the amount and location of the channel time based on the source DEV's
requirements, the loading of the superframe, and the superframe duration.

Although this method works, it requires more complexity in PNC-capable DEVs in order to support complex requests for channel time. The task group felt that it was best if the source DEV simply determined the amount of channel time it needed and requested time from the PNC rather than sending application level requirements to the PNC. In addition, the source DEV has a better understanding of the actual FER between the source and destination than does the PNC. This is particularly true in the case of 802.15.3, where the distance between the source and destination DEVs might be twice as long as the distance from either to the PNC. As the FER increases, the connection requires more time and the PNC would simply be unable to determine this unless it also kept track of lost frames at each of the destinations. Another advantage of this is that the burden of calculating the required channel time is borne by the DEV that needs to use this time.

One drawback of this method is that a really dumb DEV might simply request too much channel time because its algorithm for calculating the required channel time is not very sophisticated. However, there was no simple way to enforce good channel time requests short of specifying the actual algorithm that maps bandwidth, latency, and data buffer sizes directly into channel time and allocation repetition interval. Although there are undoubtedly many good algorithms to accomplish this, it is extremely unlikely that there would be a single one that could get the 75% agreement among the members of the task group that would be required for it to be added to the standard.

The request to create channel time to support the communication of isochronous data is initiated by higher layers in the protocol stack. The method by which the higher layers request this is unspecified in the draft standard. Once the DEV's MAC knows the amount of channel time to request, it follows the process shown in Figure 4–15 to request that time from the PNC. The source DEV begins by sending the CTRq command to the PNC. The PNC will respond to the CTRq command with a Channel Time Response command to either indicate the amount of time that will be allocated or to indicate that the reason why the request was rejected. If the request is accepted, the PNC will allocate the new channel time as soon as possible. If the request is for a sub-rate allocation, a pseudo-static allocation, or if any of the destination DEVs are in a power-save mode, then the PNC will also announce the creation of the allocation using the CTA status IE in the beacon.

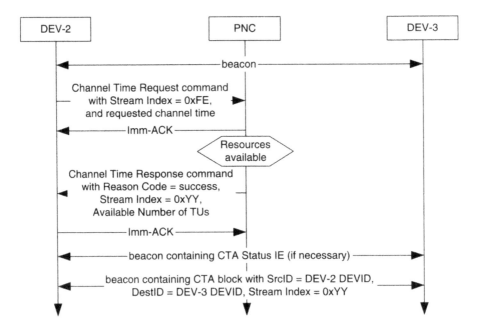

**Figure 4–15: Frame exchange for a DEV requesting
the creation of isochronous channel time**

The channel time request is based on a CTRq time unit which represents the smallest usable amount of time for the connection. The CTRq time unit is normally the length of time it takes to transmit a frame plus the time for the SIFS and the ACK, if necessary. This

> The CTRq TU can take on values from 0 to 65,535 µs, which is exactly the same as the valid range for the superframe duration. However, requesting channel time with the CTRq TU set to the maximum superframe duration will probably cause the request to be rejected.

calculation is illustrated in Figure 4–16 for frames sent using the no-ACK policy. If the DEV wants to use the Imm-ACK policy, then it needs to include the time for the frame, two SIFS duration as well as time for the Imm-ACK frame, as shown in Figure 4–17. In making this calculation, the source DEV needs to know the typical number of bits in a frame and the data rate that can be supported between the two DEVs.

The DEV also indicates in the CTRq command the minimum and maximum number of TUs that it needs for the connection. The reason for specifying the requested time in TUs is that it is the most efficient way to allocate the time. For instance, if the frame plus SIFSs and ACK duration was 0.5 ms, then it would not do any good for the PNC to allocate 1.4 ms instead of 1.0 ms because the source DEV can only send two frames during the CTA with either allocation. Although a typical TU calculation would only be for a single frame, the smallest usable time unit for a DEV might be for more than one frame, as illustrated in Figure 4–18.

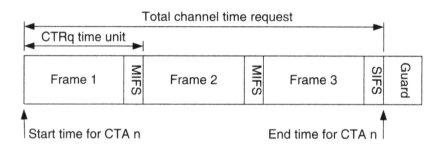

Figure 4–16: Channel time request for frames using no-ACK policy

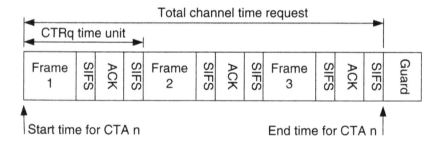

Figure 4–17: Channel time request for frames using Imm-ACK policy

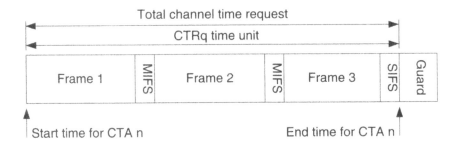

Figure 4–18: Channel time request for multiple frames using no-ACK

Fragmentation/defragmentation

A key part of any networking standard is the ability to split data up for transmission and the reassembly of the data at the receiver. Fragmentation can be done at any layer in the protocol stack. In some instances, the data that is to be sent will be fragmented by two or more of the layers in the protocol stack. IEEE Std 802.15.3 defines a method to fragment frames that would otherwise be too large to send over the PHY. The MAC may fragment a frame for a variety of reasons. The most obvious reason would be that the frame that is passed down from the higher layers is too large to transmit in a single MAC frame. Another, perhaps more subtle reason, is that the MAC may reduce the size of the frames sent over the PHY in order to improve the overall efficiency. Short frames incur a larger overhead as a percentage than do longer frames in a perfect wireless medium. However, as longer frames incur a larger FER for the same bit error ration (BER), the overall throughput of the system may be worse if larger frames are used because more of the big frames will be dropped than would the smaller frames.

In discussing fragmentation, it is important to differentiate between MAC service data units (MSDUs) and MAC protocol data units (MPDUs). The MSDU is the data "blob" that is passed between the MAC and the FCSL. The MPDU, on the other hand, is a data "blob" that is passed between the PHY and the MAC. In the 802.15.3 standard, one MSDU may be fragmented into many MPDUs, but one MPDU can never be associated with more than one MSDU.

The fragmentation method used in 802.15.3 imposes three main constraints on the DEVs. The first is that the fragments of an MSDU shall be all of the same size, except for the last one. The second requirement is that once the MSDU has been fragmented and the MAC has attempted to deliver the first MPDU, it cannot refragment the MSDU. The third major constraint is that the receiving MAC is not allowed to pass up pieces of the MSDU until it has received all of the fragments. The last requirement implies that the receiving DEV needs to have enough buffer space allocated to store the maximum size MSDU for each of the streams that it wants to be able to receive. The constant pMaxTransferUnitSize[22] is the largest MSDU that can be presented to the MAC for delivery. As this has a direct impact on the physical implementation of the DEV, this constant is PHY dependent. The 2.4 GHz PHY limits this to 2044 octets, which also the largest data payload that is allowed in an MPDU. This limit was originally 64 kbytes and it was reduced in order to save on the memory requirements. For example, to support fragmentation on eight simultaneous streams plus one asynchronous connection with a 64-kbyte pMaxTransferUnitSize would require just under 600 kbytes of RAM. This would make the cost of an 802.15.3 solution too high for the extremely cost-sensitive consumer electronics market. In addition, many other protocols use maximum packet sizes that are smaller.

The numbering of the fragments prompted quite a bit of discussion in the task group. The initial method for numbering and tracking the fragments used an MPDU number, unique for each stream, that incremented for each frame that was sent over the air. In addition, there was a fragment start and a fragment end bit in the MAC header. If the frame was one fragment long, then both the fragment start and fragment end bits would be set. If, on the other hand, one of the two fragment bits was not set, then the MPDU is one fragment of a larger frame. The first fragment would have the fragment start bit set and the fragment end bit not set. The fragments in the middle of the frame would have neither bit set, whereas the last fragment of the MSDU would have only the fragment end bit set. This method provides an unambiguous method for the receiver to correctly assemble a frame.

However, this made the logic in the receiver somewhat more difficult to implement. For example, consider the case of two MSDUs, each split into five

---

[22] The PHY constants, i.e., those that begin with a "p," are defined in Clause 11 of the standard and are cross-referenced in the last table in Clause 8 of the standard.

fragments. Suppose the receiver gets the first three fragments of the first MSDU and the last three fragments of the second MSDU. In this case, the destination would be not always be able to immediately determine that it had pieces of two different MSDUs or if it had pieces of a single MSDU. If the fragment sizes used for the MSDUs were different, then the DEV would know that there was more than one MSDU. In the case of no-ACK frames, the receiving DEV would discard all of the fragments because it did not have all of the parts of the MSDU. If the source used Imm-ACK, then the source DEV would not skip fragments unless it was going to give up on an MSDU. In this case, any gap in the MPDU number would indicate that the source DEV has given up on sending the frame, and so the destination DEV should discard any fragments that it has that have a lower number.

The only problem with the old fragmentation scheme was with the Dly-ACK mechanism. When the source uses the Dly-ACK mechanism, it may send multiple frames without waiting for individual ACKs and then request the group of ACKs at a later time. With Dly-ACK,

> At least three different voters in different letter ballots suggested fragment numbering methods that were essentially the same as the original fragmentation numbering method. However, the task group felt that they had already spent enough time discussing the issue and did not want to revisit it.

it would be possible for the receiving DEV to get the beginning of one frame and the end of another and not know they belonged to different MSDUs. The receiving DEV can hold on to all of the fragments until its internal timer determines that they should be discarded. But without knowing the MSDU from which the fragments came, the receiving DEV would not know how to group the fragments in its memory. In particular, because the fragments are only sent up as part of a complete MSDU, the MAC would typically group the fragments in memory based on the MSDU of which they were a part. Without knowing the MSDU number, it is more difficult, but not impossible, for the MAC to keep track of the memory allocations for the incoming frames.

To assist the MAC in organizing its memory, the new fragmentation process uses 3 octets (one more than before) that contain a 9-bit MSDU number, a 7-bit fragment number, and a 7-bit number that is the fragment number of what will be the last fragment. Thus, 802.15.3 has a limit of no more than 128 fragments for a single MSDU. The MSDU number is incremented on a per stream index basis just as the MPDU number was with the old method. With the new method for keeping track of fragments, the destination is able to immediately determine where to put any received frame in its memory. If the MSDU number is new, the receiver sets aside a new buffer based on the size of the received fragment and the number fragments of the MSDU, which is determined from the last fragment number. The received frame can then be put into the correct location because the destination knows the sequence number of this fragment. Although this method is not as efficient in terms of the use of the bits in the frame header, it does make the receive state machine somewhat easier to implement.

Note that the combination of the MSDU number and the fragment number is a total of 2 octets. This was done so that the field for the Dly-ACK frame, which uses these two numbers, would also be an even number of octets.

While a DEV may fragment any MSDU that it desires, only certain MAC command data units (MCDUs) may be fragmented. This restriction was introduced to simplify the implementation of the state machines at both the transmitter and the receiver. However, some of the commands are potentially much longer than the maximum frame length and so the standard allows these to be fragmented. The commands that may be fragmented are limited to the following:

a) PNC Information

b) PNC Handover Information

c) PS Set Information Response

The fragmentation and defragmentation of these commands follows the same procedure described previously for data frames.

Retransmissions and duplicate detection

In order to support reliable connections, IEEE Std 802.15.3 supports both retransmissions and duplicate detection. If the source DEV is using either Dly-ACK or Imm-ACK, it will know when one of its frames has either failed to arrive at the destination or that the source DEV was unable to receive the ACK. When the source DEV determines that it has lost a frame, it can retransmit that frame so that the destination DEV will eventually get the missing data. In the CAP, the DEV is required to use the contention-based access method (CSMA/CA) to determine when it will be able to attempt to retransmit the frame that was lost. On the other hand, in a CTA, the source DEV is not sharing the time with any other DEVs, and so it can retransmit the frame as soon as it is practical.

The other part of reliable communications is the ability of the DEV to detect when it has received a duplicate frame. Duplicate detection prevents more than one copy of an MSDU from being passed up to the higher layer protocols. An 802.15.3 DEV uses the DestID, SrcID, PNID, fragmentation control field, and stream index to determine if a frame is a duplicate of one that it has already received. In the fragmentation field, the DEV keeps track of the MSDU number and fragment number for a given stream index to watch for duplicate frames. In the case of a isochronous stream, it is easy for the DEV to determine that it has already received a frame due to the fact that the MSDU numbers will be sequential and usually will not have missing MSDU numbers. The case is quite different for an asynchronous connection because a single MSDU counter is used for potentially multiple destinations. For example, consider a DEV, DEV-1, that is sending asynchronous data to two destination DEVs, DEV-2 and DEV-3. DEV-1 sends a frame to DEV-2 with MSDU number 0 and then sends 511 frames to DEV-3 with MSDU numbers 1 through 511. If the next frame is to be sent to DEV-2, it would have MSDU number 0 and would appear as a duplicate frame. Thus, DEVs will need to use a timeout mechanism in addition to the fragmentation field to determine if it has received a duplicate asynchronous frame.

POWER MANAGEMENT

After QoS, the next most important feature of the 802.15.3 MAC is its ability to provide advanced power management features. The target market for 802.15.3 includes a large number of portable products where power usage is a critical performance measure. Although it is important to do as much as possible to reduce the active power required by the MAC and PHY, really large power savings can only come from turning off the DEV. For most DEVs, the majority of the time that the DEV is operating is spent in receive mode, waiting for something to happen in the piconet. Because of this, the transmit power requirements are secondary in importance to the receiver power requirements, especially for relatively short range PHYs like 802.11b and the 2.4 GHz PHY of 802.15.3.

Instead, it is the receiver power that dominates the battery life. Thus, an efficient power savings method needs to make sure that the receiver is only turned on when it is absolutely necessary.

The great improvement in standby time for digital cellular phones over analog phones is not due to using a different modulation or due to the digital nature of the information. The improvements in battery life are a direct result of the change from an always-on receiver for analog phones to a sometimes-on receiver for the digital phones. In digital cellular standards or specifications, e.g., GSM, TDMA, CDMA, and iDEN™, the base station will allow the phone to skip listening periods as long as it wakes up at the specified time to be able to receive an incoming call. These sleep periods can be relatively long so that the phone's radio receiver is only on a small fraction of the time. The length of time that the phone is able to sleep is limited by the latency required by the system. When a call starts, the first latency that a user experiences is measured from when they finish dialing until the other phone rings.

Normally, this latency would require less than one second for the call setup to occur, including waking up the sleeping phone. The system is able to hide the latency due to the phone waking up by sending a ring tone to the calling party even before the destination cell phone rings. The ring tone keeps the calling party from hanging up too early on the call. This allows the cell phone to be asleep for longer periods of time to save power. On the other hand, the iDEN protocol (more commonly known by the Nextel™ brand) requires that the phones listen more often in order to support the two-way walkie-talkie feature. This feature allows the caller to push a button to be "instantly" connected to the other users in the group. Although the lower latency provides some clear user advantages, it also makes it more difficult to get long standby times out of an iDEN phone.

To summarize, a good power management system that supports QoS requirements for portable devices must meet the following criteria:

a) The DEV's receiver needs to be turned off for as much time as possible.

b) The DEV should be given opportunities to turn off for >10 ms, to allow deep sleep modes (<20 μA of current).

c) The latency incurred in waking up the DEV needs to meet the application requirements.

d) DEVs that need to communicate must be awake at the same time, i.e., their wake cycles are synchronized.

e) The DEV's status as awake or asleep should be available to all members of the piconet.

f) The sleeping DEVs need to be able to preserve QoS for low data rate connections.

g) The DEVs need to be able to maintain current security information.

The 802.15.3 power management modes were designed to enable battery-powered DEVs to meet the criteria listed in a) through g) in order to get longer battery life with good QoS. In the standard, a DEV is in one of four power management modes: ACTIVE, DSPS, PSPS, or APS.[23] A DEV may be in both DSPS and PSPS modes at the same time, but it is not allowed to be in either ACTIVE mode or APS mode at the same time that it is in any other mode. If the DEV is in DSPS, PSPS, or APS modes, the DEV is said to be in a power save (PS) mode. DEVs that are in either DSPS or PSPS modes are said to be in an SPS mode because both modes offer synchronized sleep times. No matter what mode a DEV is in, it may be in one of three states, transmitting, receiving, or off. The mode that the DEV is in determines when it is required to be receiving. The DEV is never required to transmit at a particular time, with the exception of sending ACK frames when requested.

ACTIVE mode is defined as a DEV that is listening to every beacon, the CAP (if present), and all CTAs that have either the BcstID or the DEV's DEVID as the DestID. In addition, the DEV may be transmitting during the CAP or any CTAs where it is the source.

A DSPS or PSPS mode DEV, on the other hand, is only required to listen to beacons periodically. The beacons that the DEVs are required to listen to are called the wake beacons. DEVs that are in an SPS mode and are in the same PS Set share the same wake beacons, and so all of the DEVs in that PS set wake up in a synchronized fashion. This allows DEVs that share similar application requirements to be available at the same time. The latency in sending data due to a DEV waking up is reduced because DEVs that belong to an SPS set will always be

---

[23] Note that the names of the modes are always capitalized in the standard.

listening for assigned channel time at the same time. In addition, the Wake Beacon Number of all of the DSPS and PSPS DEVs is included in every beacon so that all of DEVs in the piconet will also know when to expect that the DSPS or PSPS DEV to be listening.

DEVs in APS mode, on the other hand, do not have a wake beacon that is known to either the PNC or to other DEVs in the piconet. Instead, the DEV internally determines when it is going to be awake. The only restriction for the DEV is that it needs to send a frame to the PNC at least once within an ATP to maintain its association in the piconet. For

> Timing in the piconet is relative to the beacons, and every beacon is identified by a 2-octet beacon number. The beacon number is incremented for each beacon that is sent or should have been sent (the PNC can stop sending beacons to scan other channels) and rolls over to zero once it has reached the maximum value of 65,535. The beacon number is used to indicate the time of events in the future for the members of the piconet. For example, PNC Handover IE and Piconet Parameter Change IE both use a beacon number to indicate when the event will occur. In the case of a power save DEV, the beacon number is used to indicate the next superframe that the DEV will be AWAKE.

these DEVs, the maximum latency to wake them up is their ATP, which can be quite long (in excess of 65 s). This mode is used for DEVs that wish to enter a very low power state while remaining available for occasional communications.

A summary of the four modes for 802.15.3 and their required states is given in Table 4–4.

Table 4–4: Summary of state requirements for the four modes

Superframe portion	ACTIVE	APS[a] in wake superframe	SPS[b] in wake superframe
Beacon	AWAKE	AWAKE	AWAKE
CAP	AWAKE	May SLEEP	AWAKE
CTAs with BcstID as DestID	AWAKE	May SLEEP	AWAKE
CTAs with McstID as DestID	May SLEEP	May SLEEP	May SLEEP
CTAs with DEV as SrcID or DestID	AWAKE	May SLEEP	AWAKE
All other CTAs and unassigned intervals	May SLEEP	May SLEEP	May SLEEP

[a] When the DEV in APS mode, it wakes up periodically to renew its ATP timeout.
[b] In the wake superframe for the SPS set, otherwise the SPS DEV is allowed to sleep.

In addition to the PS modes provided in the standard, DEVs in ACTIVE mode are only required to listen during portions of the superframe as indicated in Table 4–4. If the DEV is not participating in a CTA as either the source or a destination, it can shut down its transmitter and receiver to save power. This is in contrast to 802.11, where the STAs that are active are required to listen at all times. The reason that STAs in 802.11 need to listen at all times is that they do not know when a frame for which they are the destination will be sent. On the other hand, in 802.15.3, the scheduling of the time slots is given in every beacon, which allows the DEVs to know in advance when their participation will be required.

Another key difference between the power-save modes for 802.11 and 802.15.3 has to do with the network topology. In 802.11 infrastructure mode, all communications go through the AP and so there is no need for sleeping STAs to be synchronized in their wake cycles. Instead, the AP buffers all of the traffic and signals for the STA to wake up with the traffic indication map (TIM). When the STA finally wakes up, it notifies the AP, which then forwards the stored traffic to the STA. However, in 802.15.3, all communication is peer-to-peer, which greatly improves the efficiency of the network, but it also requires that the source and destination are awake at the same time. IEEE Std 802.15.3 provides a power-save mode, DSPS, that enables DEVs to synchronize their wake cycles. An added benefit of this is that these wake cycles limit the latency for data delivery, as opposed to 802.11 which provides no latency guarantee.

Common characteristics of the SPS modes

The DSPS and PSPS modes share the concepts of a wake beacon and a PS set. A PS set is a group of DEVs with similar power save requirements that share the same wake beacons. The PS Set Indices are assigned as follows:

a) PS set 0 is used for APS mode.

b) PS set 1 is used for the PSPS set (there is only one PSPS set).

c) PS sets 2-253 (0x02-0xFD) are used for DSPS sets.

d) PS set 254 (0xFE) is the unallocated DSPS set.

e) PS set 255 (0xFF) is reserved.

The APS Set Index (0) and the APS mode is a special case, as the DEVs that participate in this mode really do not belong to a set because they do not necessarily share wake beacons with any other DEV. Although the APS DEV will be in AWAKE state on a regular basis to renew its ATP timer with the PNC, none of the other DEVs in the piconet, including the PNC, will be able

> In the task group discussions, the APS mode was sometimes referred to as "cowboy mode" because the DEV gets to do pretty much anything that it wants as long as it sends one frame to the PNC every ATP. Because the PNC is already required to maintain an ATP timer for each DEV in the piconet, APS mode does not introduce a significant burden on the PNC.

to predict when this will happen. APS DEVs are assigned PS Set Index 0 in order to reuse the commands and concepts of DSPS and PSPS. In the following discussion, the references to the capabilities of DEVs belonging to a PS set does not apply to APS DEVs.

A DEV sends the PS Set Information Request command to the PNC to find out information about the SPS sets that are available. When the PNC receives this command, it replies with a PS Set Information Response command. This process is illustrated in Figure 4–19. The PS Set Information Response command contains information about the number of sets that the PNC supports and a number of PS Set Structures, one for each of the PS Sets that have members. The PNC will support at least 1 DSPS set in addition to PSPS and APS, but it may support up to a total of 252 DSPS sets.[24] The PS Set Structure contains a list of all of the DEVs that are members of that set as well as the Next Wake Beacon (i.e., the next time they will be awake) and the periodicity of those wake beacons, the Wake Beacon Interval.

---

[24] 252 SPS sets are probably more than any network or application would need to support, but there is no reason to limit the number to less than the octet that is assigned to number the sets. If an implementer is able to support that number in a cost-effective manner, there is no reason for the standard to prohibit it.

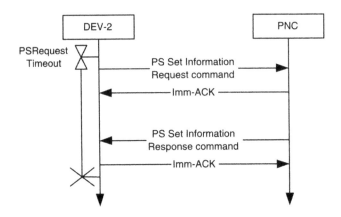

Figure 4–19: Message exchange for a DEV requesting PS set information

A DEV needs to first become a member of one or more SPS sets before it can switch to DSPS or PSPS mode. A DEV may be a member of more than one SPS set and yet not

 A DEV in APS mode is a member of PS Set 0 (the APS set). It may also be a member of other PS sets, e.g., the PSPS set or a DSPS set, while it is in APS mode. When a DEV switches between power management modes, only the APS set membership changes. A DEV remains a member of an SPS set until it is disassociated or requests to leave the set.

currently be in a PS mode. One of the reasons for this is that an SPS set exists only as long as there are DEVs in the piconet that want to use it, i.e., only as long as there are DEVs that are members of the set. If the DEV left the SPS set every time it switched from an SPS mode to ACTIVE, the PNC would be constantly deleting and creating SPS sets. However, the PNC is only able to support a finite number of SPS sets, and so it is important that unused sets are reclaimed by the PNC to be used for other DEVs. Thus, a DEV should leave an SPS set when it no longer needs to use it for power save operation.

Once the DEV has determined the characteristics of the SPS sets that are available in the piconet, it may do one of three things:

a) Join an existing SPS set (either DSPS or PSPS), because the set has parameters similar to the ones that the DEV requires.

b) Ask the PNC to create a new DSPS set with the parameters that the DEV requires. The DEV can only do this if the number of DSPS sets already allocated is less than the total number supported by the PNC.

c) Remain in ACTIVE mode because there are not any new sets available and none of the existing sets meet the needs of the DEV.

The option c), remaining in ACTIVE mode, will likely cause the DEV to use power at a greatly increased rate, to the extent that the user will notice the difference in performance. As an example, consider a digital cellular phone that supports analog roaming (i.e., a dual-mode phone). Although the phone might last five days on standby in digital mode, it might get only seven hours of standby in analog mode. As many users know from experience, the battery life is dramatically different if the phone cannot find a digital tower and instead roams to analog mode. In a consumer electronic device, this could lead the user to return the product, which incurs a large cost for both the store and the manufacturer. Unfortunately for the manufacturer, instead of returning the DEV that is the PNC, the user will likely return the portable DEV that ran out of power too fast. Thus, even though the capabilities of the PNC determine the battery life of portable DEVs in the piconet, the typical customer will not know this technical detail and will likely blame the DEV that runs out of power. It is for this reason that the standard requires that all PNC-capable DEVs support APS, PSPS, and at least one DSPS set.

Analyzing power save efficiencies

The key characteristic of any power save method is its effect on improving battery life. In most consumer applications, roughly 10% of the total available capacity of the battery can be used for wireless connectivity. Longer battery life is always preferred, but many product designers use a 10% reduction in the battery life as the point of pain above which they would not include an additional feature, such as wireless connectivity. Battery life is generally measured in mAH (the number of mA used in 1 hour) and ranges from 600 to 900 mAH for cell phones to greater than 9000 mAH (when converted to 3 V) for a laptop. This gives acceptable current usage numbers that range from 90 mAH to 900 mAH to provide the wireless connectivity.

For almost every portable consumer electronic device, most of its life is spent doing nothing; i.e., it is in a sleep state. Thus, the focus on any power savings strategy is to develop a method by which the device spends as little time awake as possible while still maintaining the responsiveness required by the user. The measure of this capability is usually referred to as the standby time of a device. A first-order approximation of the power savings provided can be calculated by summing the total current used when the DEV is in its standby mode with the total current used while the DEV is receiving, or if required, transmitting. In the case of 802.15.3, the power-save modes do not require any significant time for transmission, so the receiver current draw will tend to dominate the power drain.

The total current used during an interval, t_{int}, can be calculated using the currents required for sleep and receive, i_{sleep} and i_{rec}, respectively, and the time required for reception, t_{rec}, using the following equation:

$$\text{total current} = i_{rec}t_{rec} + i_{sleep}(t_{int} - t_{rec})$$

The above equation is a simplification. For example, it does not include any ramp-up or ramp-down time for the DEV that is required for it to prepare to receive. However, this can be accounted for by scaling the i_{rec} value appropriately. For example, if the DEV requires 5 mA for 1 ms in warm-up before it begins to use 30 mA to send a 200 μs frame, then the i_{rec} value can be changed from 30 mA to 55 mA (30 + 5*1000/200) to include the current used during warm-up. The sleep interval can also be represented as the number of superframes that the DEV is allowed to sleep, n_{sleep}, multiplied by the superframe duration, t_{sf}. Likewise, the DEV is normally only required to be in receive mode for a fraction of the superframe; i.e.,

$$t_{rec} = c_{rec}t_{sf}$$

The ratio, c_{rec}, is calculated by dividing the beacon duration by the superframe duration. A 500 octet beacon at 22 Mb/s is approximately 200 μs, whereas a typical superframe is likely 20 ms, giving a value of $c_{rec} = 0.01$. Likewise, the current for the sleep mode can be represented with a similar equation:

$$i_{sleep} = c_{sleep}i_{rec}$$

Typical values for the ratio of receive current to sleep current, c_{sleep}, range from 0.001 to 0.01.

Using the preceeding equations, the total current can be represented as:

$$\text{total current} = i_{rec} c_{rec} t_{sf} + i_{rec} c_{sleep} (n_{sf} t_{sf} - c_{rec} t_{sf})$$

which can then be simplified to:

$$\text{total current} = i_{rec} t_{sf} (c_{sleep} n_{sf} + c_{rec} - c_{rec} c_{sleep})$$

For a typical radio, the values of c_{rec} and c_{sleep} will likely be small, so that the last term can be deleted as it will be much smaller than the other two terms in the parentheses. Finally, the power savings that is achieved by a method that allows a DEV to sleep for a number of superframes compared with waking up for every superframe, i.e., $n_{sf} = 1$, can be calculated from:

$$\text{increase in battery life} = \frac{c_{sleep} n_{sf} + c_{rec}}{c_{sleep} + c_{rec}}$$

Using values of 0.01 for c_{rec} and c_{sleep}, the increase in battery life is plotted in Figure 4–20. As expected from the simplified equation, the increase in battery life is linearly proportional to the increase in the interval between wake beacons. This is intuitive because the power usage of the DEV will rise roughly linearly with the number of times it needs to be in receive mode, because this dominates the overall power usage.

Doubling the Wake Beacon Interval (802.15.3 only allows powers of 2 for the Wake Beacon Interval) will then lead to a significant improvement in the overall battery life. This result is graphed in Figure 4–21 using 0.01 as the value for c_{rec} and c_{sleep}. As the graph shows, once the Wake Beacon Interval is increased beyond 16 (i.e., 2^4), doubling the Wake Beacon Interval doubles the battery life. Even for relatively short Wake Beacon Intervals, the improvement is at least a factor of 1.6. This result provides the primary motivation in 802.15.3's power save strategy: Allow the DEVs to sleep for as long as they need to support the requirements of their application.

The preceding calculations do not take into account any sub-rate traffic that the DEV may need to either send or receive while it is in sleep mode. If the time spent on this traffic is significant, then the improvement from skipping wake beacons will be less pronounced. In this case, the sub-rate interval would have a stronger impact on the overall battery life. In any case, using the best sub-rate or Wake Beacon Interval is key to improving the battery life of the DEV.

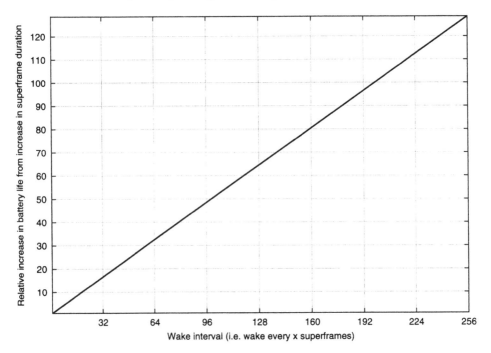

Figure 4–20:   Increase in battery life as a function of the number of superframes skipped

Switching PM modes

A DEV changes its PM mode by sending the PM Mode Change command to the PNC with the PM Mode field set to the appropriate value. If the DEV is making a valid request, the PNC will set the bit for the DEV in the appropriate PS Status IE in the next beacon. When the DEV receives a beacon with its bits set, it knows that it can change its PM mode. The PS DEV needs to wait for this indication from the PNC so that other members of the piconet can be informed that the PS

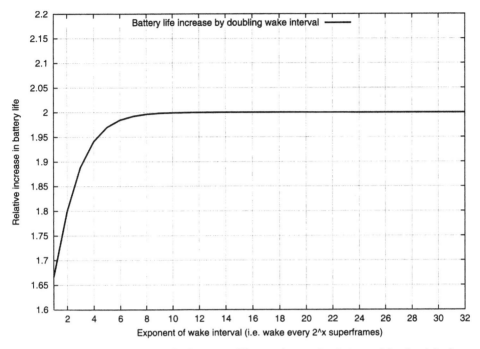

Figure 4–21: Increase in battery life as the wake interval is doubled

DEV will not be listening to every beacon. If the PM Mode field is set to "SPS," the PNC will register the DEV as participating in all of the SPS sets that it has joined. For example, if the DEV is a member of the PSPS set and two other DSPS sets when it sends the PM Mode Change command with the PM Mode field set to "SPS," the DEV will be required to listen to the wake beacons of all three sets. Figure 4–22 illustrates the frame exchange for a DEV changing from ACTIVE mode to DSPS, PSPS, or both.

Normally, the PNC will always grant a request to change the DEV's PM mode. The only reason for the PNC to reject this request is if the DEV is requesting to change to SPS mode and it is not currently a member of a valid SPS set (i.e., a set with a value of 1–253). If the DEV is not a member of a valid SPS set, the PNC will do nothing with the beacon and the DEV will still be required to be in ACTIVE mode and to listen to every beacon.

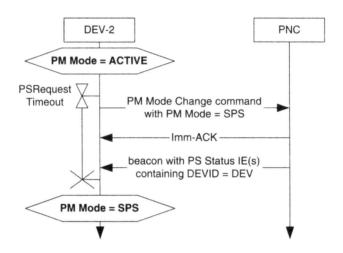

Figure 4–22: DEV changing from ACTIVE mode to DSPS, PSPS, or both

A DEV changes from any PS mode (either APS or SPS) to ACTIVE mode by sending the PM Mode Change command to the PNC with the PM Mode field set to "ACTIVE". This frame exchange is illustrated in Figure 4–23. The DEV will consider itself to be in ACTIVE mode once it has sent the PM Mode Change command to the PNC, even if the DEV doesn't receive an ACK. The reason for this is that the DEV will not harm the piconet operations if it begins listening to every beacon. In additon, there is no reason for the PNC to refuse the request, as there is in the case when the DEV switches from ACTIVE to SPS mode.

Even if the ACK was not received by the DEV, it is still possible that the PNC correctly received the PM Mode Change command. If the DEV does not see a change to its PS status in the beacon after requesting the change to ACTIVE mode, it can re-send the PM Mode Change command to the PNC. If the DEV was in SPS mode, it will still remain a member of its SPS sets after it changes to ACTIVE mode. A DEV retains its SPS set membership until it explicitly requests to leave the set by sending the proper the SPS Configuration Request command to the PNC.

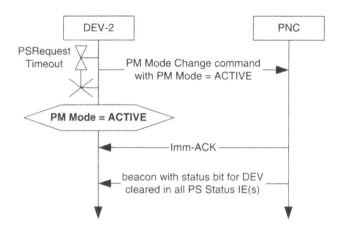

Figure 4–23: DEV changing from any PS mode to ACTIVE

A DEV changes from ACTIVE mode to APS mode by sending the PM Mode Change command to the PNC with the PM Mode field set to "APS," as illustrated in Figure 4–24. The DEV considers itself to be in APS mode when it receives the ACK to its PM Mode Change command. When a DEV changes from any other PM mode to APS mode, all of the streams and asynchronous allocations where the DEV is either the source or destination will be terminated by the PNC.

If the PM Mode field in the PM Mode Change command is set to a value that is the same as the current mode of the DEV, then the DEV is not requesting a change in its PM mode. The PNC takes no action in response to this command and the PS Status IE(s) remain the same after it has received the command.

The PS Status IEs are used to keep the DEVs in the piconet informed of the current power save status of every DEV in the piconet. The PNC places a PS Status IE in the beacon for each PS Set that has a DEV in a PS mode. The PS Status IE contains the PS Set Index, the set's Next Awake Beacon, and a bitmap that indicates the DEVs in the set that are currently in a PS mode. For example, if there is a DEV in APS mode, the PNC places a PS Status IE in the beacon with PS Set Index set to 0 (the APS set) and the bit that corresponds to that DEV set to 1. Likewise, if a DEV has joined the PSPS set and two DSPS sets prior to changing to SPS mode, the PNC will place three PS Status IEs in the beacon, one for set 1 (the PSPS set) and one for each of the DSPS sets that the DEV has joined. The PNC only places one PS Status IE per active PS set in the beacon.

Because the PS Status IE is in every beacon when there is a DEV in a PS mode, all of the DEVs in the piconet will know which DEVs are in a power-save mode. In addition, the DEVs in the piconet use the Next Awake Beacon field to determine the beacon number of the superframe when the power save DEVs will be awake to listen for traffic. This allows a DEV to determine when it will be able to connect with DEVs that are in a PS mode. The exception to this, of course, is for DEVs that are in APS mode. In this case, the other DEVs in the piconet will be informed that the DEV is in APS mode, but they will not know when that DEV will be awake to listen for traffic.

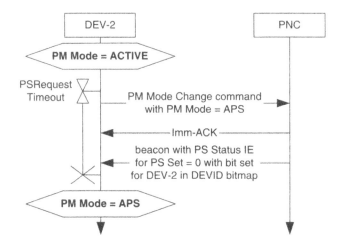

Figure 4–24:   DEV changing from ACTIVE mode to APS

Managing SPS sets

Only DSPS sets can be created by a DEV because the PSPS set belongs to the current PNC. If a DEV decides to create a new DSPS set because none of the existing sets meet its needs, it sends an SPS Configuration Request command to the PNC with the PS Set Index equal to the Unallocated PS set. The Unallocated PS set has a value of 0xFE, which is the same value as the unassociated ID used in the association process. The DEV also sets the Operation Type field to indicate "join" and requests a specific beacon interval with the Wake Beacon Interval field.

The PNC can only accept or reject the request; it cannot modify the requested Wake Beacon Interval.

Although the PNC cannot modify the interval of the wake beacons, it can determine when the first wake beacon will occur. This offers the PNC the opportunity to spread out the wake beacons of the DSPS sets so that all of the DEVs in the piconet are not waking up at the same time. If the PNC accepts the request to create a new PS set, it assigns a number, the PS Set Index, that will be used to identify the set. The PNC then responds to the requesting DEV with the SPS Configuration Response

> The 802.15.3 standard uses 0xFE for the unassociated ID and the unallocated PS set. This number was selected because 0xFF had already been selected to be used for the broadcast ID. Traditionally, the broadcast address is one that has all of the bits set in some portion of the address. The next available ID, counting down from 0xFF, was 0xFE and this was selected as the unassociated ID. The task group then adopted this number for the DSPS set creation process because the PNC was assigning an ID, just as it does in the association process.

command that includes the new PS Set Index and the Next Wake Beacon for the new PS set. The frame exchange for creating a new DSPS set is illustrated in Figure 4–25.

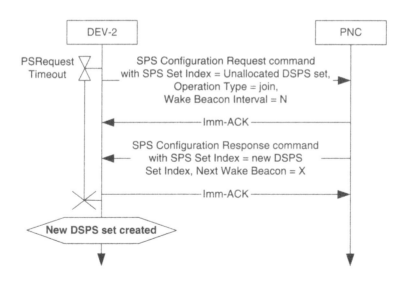

Figure 4–25: Frame exchange for creating a new DSPS set

Once the DSPS set has been created, the PNC keeps the Next Wake Beacon for that set updated and broadcasts this number in the PS Status IE in the beacon when there are DEVs in that DSPS set that are in DSPS mode. The PNC also keeps track of the Wake Beacon Interval that was requested when the set was created. Although the PNC is allowed to change the Next Wake Beacon, i.e., it is allowed to change the locations of the DSPS set's wake beacons relative to other DSPS sets, the Wake Beacon Interval is kept constant for the life of the PS set. Neither the PNC nor a DEV is allowed to change the interval while the PS set exists.

> The original draft of DSPS mode (at the time called EPS mode) allowed the DEV that created of the DSPS to modify the parameters of the set. However, if this DEV left the set, it wasn't clear which of the remaining DEVs in the set should be given the power to modify the set parameters. Instead of trying to create a line of succession of DEVs that would have the right to modify the set parameters, the task group decided that no DEV should be able to modify the set parameters. In one of the last drafts this was modified slightly. The PNC can change the Next Wake Beacon, but Wake Beacon Interval would not be changed while the set was in use.

If the DSPS sets share some of their wake beacons, then in those superframes the total duration of the CTAs requested by all of the DEVs may exceed the superframe duration. This problem is referred to as differential overloading. If every superframe is full, i.e., there is not any additional time available for CTAs in any superframe, then the PNC has a problem of superframe overloading. However, DEVs in a DSPS mode will often request sub-rate allocations that are aligned with their wake beacons to save power. In non-wake beacons, the allocations will not exist and the superframe will be under-utilized. On the other hand, when the wake beacons of multiple DSPS sets overlap, then that particular superframe may be overloaded due to the additional sub-rate slots. However, because the PNC is able to arrange the wake beacons of the DSPS sets, it is able to minimize the potential of differential overloading of the superframe.

Once a set has been created, a DEV may join the set so that it can use it for power save purposes. To join the set, the DEV sends the SPS Configuration Request command to the PNC with the PS Set Index field equal to the index of the set that the DEV wishes to join and the Operation Type set to "join." As the DEV is not allowed to change the set parameters of an existing set, the Wake Beacon Interval field is ignored by the PNC. The PNC can only reject the request if the PS Set Index in the request does not exist. If the DEV is already a member of the PS set, the PNC takes no additional action other than to send a response. Otherwise, the PNC will add the DEV to the PS set as a member. Regardless of the result of the request, the PNC sends the SPS Configuration Response command to the DEV with the appropriate Reason Code to indicate if the request failed or was successful. The frame exchange for joining an existing PS set is illustrated in Figure 4–26.

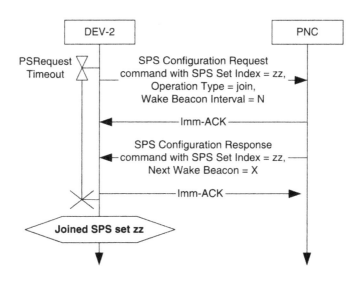

Figure 4–26:   Frame exchange for joining an existing SPS set

When the DEV no longer needs to use the PS set, it sends the SPS Configuration Request command to the PNC with the Operation Type set to "leave." All of the other parameters in the command are ignored by the PNC. The PNC will always accept a request by a DEV to leave a PS set and the PNC does not send the SPS Configuration Response command to the DEV. Instead, the Imm-ACK sent in response to the SPS Configuration Request command serves as the notification that the PNC will remove the DEV from the PS set. The frame exchange for leaving a PS set is illustrated in Figure 4–27. When the last DEV has left a DSPS set, the set is terminated by the PNC. Note that the PSPS set is never terminated because it is the PNC's set, and so it exists as long as the piconet exists.

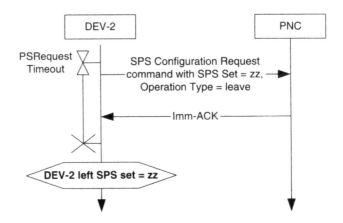

Figure 4–27: Frame exchange for leaving an SPS set

DSPS mode

DSPS mode is designed to give extremely good power savings while allowing the DEVs to control the latency of data transmission in sleeping DEVs. DSPS mode also provides a method by which DEVs that share common requirements for power

 The original version of DSPS mode had the name extended power save (EPS) mode. This was later renamed synchronized power save (SPS) mode before the task group settled on DSPS mode.

savings are able wake up at the same time to communicate. The DSPS mode was the basis for the PSPS mode, which trades away some of the power savings in DSPS mode in order to provide a piconet-wide wake beacon for participating DEVs.

Once a DEV has joined a DSPS set (see "Managing SPS sets" on page 114), it switches between DSPS and ACTIVE mode depending on the type of traffic that it is handling. If the DEV has infrequent traffic, it can remain in DSPS mode and use a sub-rate allocation to maintain communications with another DEV. If both the source and destination need to save power, then they would both join the same DSPS set. When a DEV changes from ACTIVE to DSPS, the PNC will terminate all streams for which the DEV is the source or destination that are not using a DSPS Wake Beacon Interval. The reason for this is that the DSPS DEV will not be listening to every beacon and hence will not be able to transmit or receive during those CTAs. However, when a DEV changes from DSPS to ACTIVE mode, the PNC does not terminate any streams, regardless of the type of allocation.

When a DEV is in DSPS mode, there is no guarantee that it will be able to receive either broadcast or multicast frames. Unless the request for the broadcast or multicast allocation was made with a DSPS Set Index, the PNC will not attempt to align the allocation to the DSPS set's wake beacons. Thus, if a DEV needs to have all of the DEVs in the piconet receive the frame, it will have to repeat the message in at least one wake beacon of each of the PS sets. However, broadcast delivery is not guaranteed anyway, so the source DEV can only provide the opportunity for the DEVs to receive the broadcast frame; it cannot guarantee that they will receive it. The only method to ensure that a DEV receives a frame is to send it as a directed frame with the ACK policy set to either Imm-ACK or Dly-ACK. Likewise, if a DEV requires the opportunity to receive all broadcast and multicast frames, it should not use a PS mode.

Allocating channel time for DSPS DEVs

Although DSPS provides clear advantages in power savings for DEVs in an 802.15.3 piconet, another key advantage comes from the ability to synchronize low power, low data-rate applications while maintaining the required QoS. The DEVs accomplish this by allocating sub-rate streams that are aligned with the DSPS set's wake beacons. A sub-rate stream is one where the allocation occurs only once every N beacons. This type of allocation allows the DEVs to maintain a connection with a defined latency and guaranteed bandwidth. Applications that could use this include remote control functionality for A/V equipment and cordless phones (to provide a defined latency to wake up the handset). Another application is a remote security camera that is triggered on movement. It would maintain a low-rate connection to verify its status as active, but it would then switch to a high-rate stream when it had video to transmit due to activity in its area.

The DSPS sub-rate allocations are aligned with the DSPS set's wake superframes, so their allocation interval, the CTA Rate Factor, needs to be either greater than or equal to the DSPS set's Wake Beacon Interval. For example, if a DSPS set uses a Wake Beacon Interval of 8, then a DEV could request DSPS channel time with a CTA Rate Factor of 8, 16, 32, 64, and so on. This is illustrated in Figure 4–28 for various Wake Beacon Intervals and CTA Rate Factors. In order to minimize the complexity required for sub-rate allocations and DSPS sets, the PNC is only required to support CTA Rate Factors and Wake Beacon Intervals that are powers of 2 (i.e., 2, 4, 8, 16, ...). In case 1 in Figure 4–28, the Wake Beacon Interval is equal to the CTA Rate Factor and so the CTAs occur in every wake superframe. In case 2, however, the CTA Rate Factor is twice that of the Wake Beacon Interval and so the CTAs occur only in every other wake superframe. Some combinations of CTA Rate Factor and Wake Beacon Interval, however, are not allowed. For example, a CTA Rate Factor of 2 would not be allowed for a Wake Beacon Interval of 8. In this case, the CTAs would occur more often than wake beacons and so some of the allocations would occur in superframes when the DEV is not going to be awake.

The DEV that is creating or modifying a stream uses the PM CTRq Type field of the Channel Time Request command to indicate if the requested allocation needs to be aligned to a DSPS wake beacon. If a DEV requests a DSPS allocation, the PNC will align the channel time with the wake beacons of the DSPS set indicated in the request. If the interval for the sub-rate allocation is greater than the Wake Beacon Interval, then the PNC has flexibility in assigning the CTAs, as illustrated in Figure 4–29.

An ACTIVE mode sub-rate allocation, on the other hand, can occur in any beacon. This gives the PNC the flexibility to spread out the loading of the sub-rate allocations among the superframes. An example of spreading out the sub-rate allocations is illustrated in Figure 4–30. In this example, eight DEVs have joined a DSPS set with a Wake Beacon Interval equal to two. They have also requested sub-rate allocations with a CTA Rate Factor equal to eight. The PNC can then pair the requests and start a different pair every other superframe, reducing a potential load of eight CTAs in a superframe to only two CTAs in a superframe.

Sub-rate allocations, however, do not necessarily have the same priority as super-rate allocations. Because the sub-rate allocations can occasionally align to overload the superframe, the PNC is allowed to skip the allocations and can then take one of three actions to handle the differential overloading problem for DSPS sub-rate allocations:

a) Take no other action, and continue the allocations if time is available in future superframes.

b) Terminate the stream because it is unable to service the request reliably.

c) Set the DEV's bit in the continued wake beacon (CWB) IE, and allocate the CTA in one of the three following superframes.

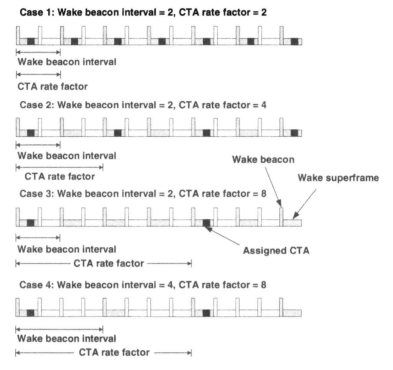

Figure 4–28: Illustration of the relationship between Wake Beacon Intervals and sub-rate CTAs

When a DSPS DEV sees its bit set in the CWB IE, it knows that the PNC was unable to provide one of its allocations in the current wake superframe. The DEV will then listen to the next beacon to see if the allocation appears in that beacon. The PNC can again set the DEV's bit in the CWB IE to keep the DEV awake for another beacon. This process, referred to as chaining wake beacons, allows the PNC to handle infrequent differential overloading problems in the piconet. However, the PNC is only allowed to set the DEV's bit in the CWB IE for up to three consecutive beacons. In general, the PNC should avoid this method as much as possible because it will have a negative impact on the battery life of the DEV.

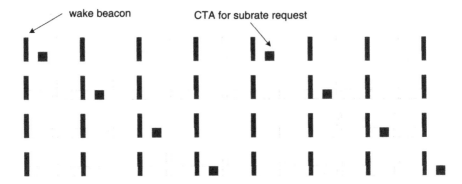

Figure 4–29: Equivalent CTAs for Wake Beacon Interval = 2 and CTA Rate Factor = 8

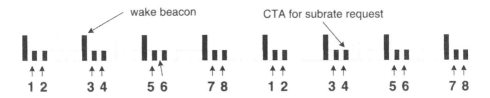

Figure 4–30: Minimum superframe loading of CTAs for 8 DEVs requesting Wake Beacon Interval = 2 and CTA Rate Factor = 8

One of the requirements of any power save technique is that it provides the ability of other DEVs to "wake up" a DEV that is in a power-save mode. In IEEE Std 802.15.3, the wake up process is linked with the channel time request process. This is a natural linkage because

> An ACTIVE mode CTA is one where the time allocation is not intentionally aligned with a DSPS set's wake beacons. A DEV requests an ACTIVE mode CTA by setting the PM CTRq Type bit in the Channel Time Request command to be zero.

the reason that one DEV needs to wake up another DEV is so that it can begin communications, and hence, it needs to allocate channel time for this to happen. The wake up process begins when a DEV requests an ACTIVE mode CTA with a target DEV that is in DSPS mode. If there is time available for the allocation, the

PNC will grant the request and send a response to the requesting DEV indicating that the request has been granted, but that the destination DEV is in a power-save mode. The PNC will then set the bit corresponding to the DEV in the pending channel time map (PCTM) IE to inform the sleeping DEV that it has a new allocation, not aligned to its wake beacon that is pending.

When a DEV in DSPS mode receives a beacon with the PCTM IE that has the DEV's bit set, the DEV will send a PM Mode Change command to the PNC. If the DEV is going to keep the allocation, it sets the PM Mode in the command to ACTIVE and the PNC will then begin to place the appropriate CTA blocks in the beacon. If, however, the DEV does not want to keep the allocation, it responds to the PNC with a PM Mode Change command with the PM Mode set to DSPS. The PNC will then terminate the stream using the standard stream termination procedure (see "Asynchronous data" on page 87).

The process of creating ACTIVE mode channel time with a DSPS DEV is illustrated in Figure 4–31.

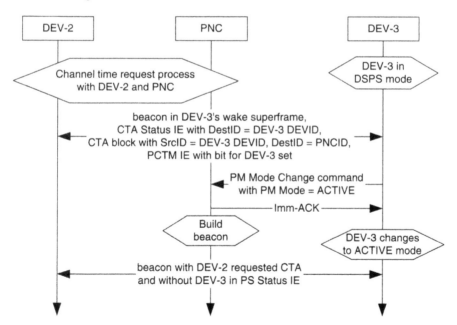

Figure 4–31: Channel time request causing a DSPS mode DEV switch to ACTIVE mode

PSPS mode

PSPS mode is very similar to DSPS mode. The goal of PSPS mode, however, is very different than the goal of DSPS, and so it has some characteristics that are also different. Although the goal of DSPS is to provide optimum power savings while maintaining QoS, the goal of PSPS is to provide a method for ensuring that sleeping DEVs will be awake during superframes when important announcements might be made. PSPS mode does not use a regular Wake Beacon Interval. Instead, the PNC is allowed to change the Wake Beacon Interval in between every wake beacon. Thus, the two key differences between PSPS and DSPS is that in PSPS mode it is the PNC that determines the Wake Beacon Interval and that the PNC can change that interval in every wake beacon.

When a DEV joins the PSPS set and switches to PSPS mode, it is required to listen to all system wake beacons, i.e., the wake beacons for PS set 1. If it misses receiving a system wake beacon, the DEV is required to continue to listen for beacons until it successfully receives a beacon. One reason for this is that if the DEV does receive a beacon, it will not know the beacon number of the next system wake beacon. A DSPS DEV, on the other hand, does not need to

> The idea for a system-wide wake beacon was first proposed at the IEEE 802 Plenary meeting in St. Louis, MO, in March 2002, as a way to handle broadcast transmission for sleeping DEVs. The idea was that power-save mode DEVs that wanted to hear broadcast traffic would be awake during the system wide wake beacon. Instead of adopting this approach, the group decided not to attempt to guarantee that all DEVs in the piconet would have the opportunity to receive broadcast traffic. If a DEV needs to hear this traffic, it cannot enter a power-save mode. The idea of a system wake beacon appeared again when the PSPS mode was proposed at the IEEE 802 Plenary meeting in Vancouver, BC, in July 2002. This mode was named PSAVE mode in its proposal but was changed to PSPS when it first appeared in a draft.

listen to subsequent beacons because it knows that the Wake Beacon Interval is fixed and so it can calculate the beacon number of the next DSPS wake beacon.

In PSPS mode, the Wake Beacon Interval is not constant, and so it is not possible for the PNC to align sub-rate allocations with the system wake beacon. Even if the PNC does allocate the first CTA during a system wake beacon, there is no guarantee that this will happen in the future. In general, a PSPS mode DEV will burn much more power than an equivalent DSPS mode DEV. One reason is that the PSPS mode DEV will listen to more beacons to make up for beacons that were lost. It will also use more power because the DEV will have to wake up once to listen to the system wake beacon and will have to wake up again in a different superframe to receive frames during its sub-rate allocation. If the system wake beacon is the same interval as the sub-rate allocation, this would have the effect of almost doubling the power drain on the battery due to the DEV being required to listen to an extra beacon. The effect is less pronounced if the sub-rate occurs much less frequently than the system wake beacon, but in all cases, using a DSPS set would result in improved power efficiency.

A final issue that degrades the power save capabilities of PSPS mode is that the PNC and not the DEV determines the interval between system wake beacons. A DEV can send a suggested interval to the PNC, but the PNC is not required to honor this request. For example, suppose three PSPS DEVs request Wake Beacon Intervals of 128 while two PSPS DEVs request Wake Beacon Intervals of 16. If the PNC uses 128 as the system Wake Beacon Interval, then the DEVs with a need for an interval of 16 could suffer from higher latency. However, if the PNC uses an interval of 16, then the other three DEVs will see a dramatic increase in their power usage as compared with an interval of 128. Furthermore, the PNC may have a policy of allowing system Wake Beacon Intervals of no more than 8 to ensure a fast response time for beacon announcements. In this case, none of the DEVs would get their requested beacon interval and the power usage for all of the PSPS DEVs would increase. As shown in Figure 4–21, decreasing the Wake Beacon Interval from 16 to 8 could increase the DEVs impact on the battery life by a factor of 2.

When a DEV switches from ACTIVE mode to PSPS mode, the PNC will terminate any super-rate isochronous streams that have the DEV as the destination. If the PNC did not do this, then the PSPS DEV would either miss the CTAs that occur in every beacon or it would listen to every beacon, which is the same as being in ACTIVE mode. The PNC will allocate asynchronous CTAs that have the PSPS DEV as the destination during the system wake beacons so that the sleeping DEV has a chance to receive this traffic. If the PNC is using the group allocation method and some of the DEVs in the group are in an SPS mode, the PNC will attempt to schedule as many of the asynchronous allocations as possible during the system wake superframe, but it is not required to put any of the allocations in that superframe. DEVs that have broadcast or multicast allocations can use the system wake beacon to reach all of the DEVs that are in PSPS mode.

The main reason to use PSPS mode is to provide defined beacons in which certain announcements will be made. For example, if there are DEVs in PSPS mode, then the Piconet Parameter Change IEs will occur at least once in a system wake beacon and in at least three consecutive beacons that follow the system wake beacon. This improves the probability that the PSPS DEVs will find out about the upcoming change in one of the key parameters of the piconet. If a DEV misses an important change announcement, e.g., a channel change, it will have to take action to reacquire the piconet. This does not add complexity because searching for and acquiring synchronization with a wireless network is an essential function of all portable wireless devices. The uncertainty of the wireless medium requires the designer to always plan for the inevitable condition where synchronization is lost with the desired network.

APS mode

APS mode was introduced to complement the other two power-save modes APS has two important power saving capabilities that are different from either DSPS or PSPS. The first capability compensates for the limited number of DSPS sets. Hardware limitations, e.g., storage space in the MAC and processing power, imply that the number of DSPS sets will be limited in a PNC, possibly to a relatively small num-

APS was in one of the early drafts of the standard, but was removed because the initial version was broken. A simplified version of APS that fixed the earlier problems was added during Sponsor ballot based on the argument that it added almost no complexity to the PNC. This version was named HIBER-NATE, but the technical editor felt that the name was too long for it to be displayed nicely in the standard. Because the other power-save modes were already using PS as a part of their name, APS selected to be the name for the final version of this power-save mode.

ber, e.g., four. Once those sets are all defined, new DEVs that have different power save requirements, e.g., low sensitivity to latency but high sensitivity to power usage, might be unable to match the desired power save performance to the existing sets. APS mode allows the DEV to choose any Wake Beacon Interval that it requires without increasing the demands on the PNC.

The second capability provided by APS is that it does not require that the sleeping DEV keep an accurate estimate of the superframe timing. In PSPS and DSPS modes, the DEVs are required to wake up for specific beacon numbers and will need to determine an estimate of the drift between the PNC's clock and the DEV's clock so that the sleeping DEV will wake up at the right time for the wake beacon. In APS mode, however, the sleeping DEV can wake up at any time and listen to the next beacon that arrives.

There are, however, three drawbacks to APS mode. The first is that a DEV in APS mode is not allowed to have any isochronous streams or asynchronous allocations while in APS mode. When the DEV switches from any other PM mode to APS mode, the PNC will terminate all isochronous and asynchronous allocations that have the DEV as either the destination or the source. In the case of a group asynchronous allocation, the PNC will remove the DEV from the target list in the CTRqB without changing the total requested allocation.

The second drawback to APS mode is that other DEVs in the piconet do not know the superframes during which the APS DEV will be listening. Thus, when a DEV requests channel time with an APS DEV, the requesting DEV will not know for sure when the APS DEV will respond to the PNC so that the allocation can be put in the superframe. The requesting DEV can be aware of the ATP value for the APS DEV because that value is included in the PNC Information command, which is periodically broadcast by the PNC. Thus, the requesting DEV will be able to place an upper limit on the length of time that it

> IEEE Std 802.11 provides power-save modes in a manner that is similar to APS mode in 802.15.3. In 802.11, the AP stores all frames intended for sleeping STAs and forwards them to the sleeping STA when it wakes up. The AP uses the traffic indication map (TIM) to inform a sleeping STA that it has frames that need to be delivered. Thus, the power-save modes in 802.11 only work in infrastructure mode and not in ad-hoc mode. This procedure works well because the AP is normally a stationary device that has access to AC power. However, this approach increases the memory and cost for the AP. It also introduces an unpredictable latency in the delivery of the frame, which is unacceptable for applications that require QoS.

will take the APS DEV to respond, but that could end up being a very long time (the longest ATP allowed is approximately 65 seconds).

The third drawback of APS mode is that the wake beacons of a pair of DEVs in APS mode are not synchronized. Thus, if the DEVs need to send occasional traffic to each other, the latency in sending the traffic can be quite large. For example, if one of the DEVs needs to communicate with the other DEV, it first needs to switch to ACTIVE mode and then request channel time from the PNC. At this point, the other APS DEV may have just finished the process of renewing its ATP and will not be listening to beacons for quite some time.

A DEV can use APS mode without first joining the APS set. Instead, the DEV simply uses the PM Mode Change command to switch to APS mode (see "Switching PM modes" on page 110). The method used to wake up a DEV in APS mode is the same as for DSPS. The requesting DEV sends a Channel Time Request command to the PNC, and the PNC sets the appropriate bit in the PCTM IE if the request will be granted. When the APS DEV responds to the PCTM IE with the PM Mode Change command, it can either accept the allocation by switching to ACTIVE mode or refuse the allocation by staying in APS mode.

CHANGING PICONET PARAMETERS

DEVs that are members of the piconet synchronize with the PNC based on the PNID, the BSID, the PHY channel, and the timing of the beacon. Whenever the PNC changes one of these key characteristics of the piconet, it uses the Piconet Parameter Change IE to communicate the upcoming change to the members of the piconet. The PNC is required to repeat this announcement in at least mMinBeaconInfoRepeat consecutive beacons. The parameter mMinBeaconInfoRepeat is used as a basis for other announcements and has a value of 4 in the standard. The significance of this number is that it is the same as the number of beacons that a DEV with a pseudo-static allocation can miss before before it must stop transmitting until it sees another beacon. This repetition ensures that DEVs that are the source of pseudo-static allocations will either hear at least one announcement when the PNC is changing the size of the superframe or the location of the beacon or that they will have to stop transmitting until the DEVs receive a beacon.

The Piconet Parameter Change IE indicates the parameter that the PNC will be changing and the beacon number of the superframe in which the change takes effect. The PNC is only allowed to change one piconet parameter at a time; i.e., it is not allowed to indicate a change to the superframe size while it is also indicating an impending channel change. In addition, the PNC is not allowed to change the locations of any pseudo-static CTAs once it is has indicated that it will change one of the piconet parameters.

During the parameter change, there may be DEVs in the piconet that are operating in a power-save mode. If there are any DEVs in PSPS mode, the PNC is required to place at least one occurrence of the Piconet Parameter Change IE in a system wake beacon and in at least three consecutive beacons that follow the system wake beacon. All of the DEVs in PSPS mode are required to be awake to receive the system wake beacon, and if the PSPS DEVs miss it, they are required to be awake for subsequent beacon transmissions until a beacon is received. Although it is possible that a PSPS DEV might miss all four beacon announcements, the probability of this is pretty low.

The PNC does not have to align the piconet parameter change announcements with DSPS wake beacons, although the PNC could continue the announcements beyond the minimum of four repetitions so that they occur in DSPS wake beacons as well. It is not possible for the PNC to align the announcements with the wake times of APS DEVs because each APS DEV can wake up at a different time and that time is not known by the PNC. If a DEV in any of the three power-save modes hears the announcement, it is not required to switch to ACTIVE mode. The sleeping DEVs can calculate the effect of the change based on the information in the Piconet Parameter Change IE and adjust their settings such that when they next wake up, they will still be synchronized with the piconet. Thus, there is no reason to force the sleeping DEVs to ACTIVE mode when a piconet parameter changes. On the other hand, requiring the sleeping DEVs to switch to ACTIVE mode would increase the command traffic right before and right after the parameter change as well as needlessly increasing their power usage.

Beacon announcements

From time to time, the PNC is required to announce changes that occur in the piconet. Some of the changes apply to every DEV in the piconet, whereas others only apply to a single DEV. In general, the PNC is required to repeat these announcements so that most of the DEVs in the piconet will have the opportunity to hear the announcements. However, because these announcements are broadcast in nature, there is no guarantee that a DEV will receive the announcement. To handle this case, the content in these announcements can also be retrieved from the PNC using a command, usually the Probe Request command. Some of the beacon announcements are not repeated, but rather appear whenever they are needed. IEs that are put in the beacon on an "as needed" basis include the following:

• Application Specific

- Pending Channel Time Map (PCTM)
- PS Status
- Continued Wake Beacon (CWB)

Table 4–5 has a listing of the announcements that a PNC might make in a beacon for which there is a minimum number of times that they will be repeated. The table lists the announcements, the minimum repetition for the announcements, and if the announcement is intended for a single DEV or if it is intended for all of the DEVs in the piconet.

Table 4–5: Beacon announcements

Information element	Repetition	Intended for
DEV Association	mMinBeaconRepeatInfo	All DEVs
PNC Shutdown	mMinBeaconRepeatInfo	All DEVs
Piconet Parameter Change	mMinBeaconRepeatInfo	All DEVs
PNC handover	mMinBeaconRepeatInfo	All DEVs
CTA status	mMinBeaconRepeatInfo	DestID

The rules for beacon announcements are simple. All announcements are made in at least mMinBeaconRepeatInfo consecutive beacons beginning with any beacon. If the intended recipient of an IE is all DEVs, then if any DEV is in PSPS mode, the IE announcement is made in a system wake beacon and in at least mMinBeaconInfoRepeat-1 consecutive beacons following the system wake beacon.

If the intended recipient of an IE is a single DEV that is in SPS mode, the following rules apply:

- If the DEV is in PSPS mode, the IE announcement is made in a system wake beacon and in at least mMinBeaconInfoRepeat-1 consecutive beacons following the system wake beacon.
- If the DEV is in DSPS mode, the IE announcement is made in one of the DEV's DSPS set wake beacons and in at least mMinBeaconInfoRepeat-1 consecutive beacons following the DEV's DSPS set wake beacon.

A CTA Status IE is considered to be intended for all DEVs if the DestID contained in that IE is the BcstID or McstID. Otherwise the CTA Status IE is intended for the DEV defined by DestID.

There is a special exception for the PNC Shutdown IE because the request to turn off the PNC may have a timeout that is too short to support normal beacon announcements. The shutdown announcement is a courtesy to the DEVs in the piconet. From a DEV's point of view, the piconet may disappear at any time without a shutdown announcement as the DEV can move out of range of the PNC in its normal operation. Thus, the PNC will attempt to send shutdown information, but when it needs to stop, it will cease operation.

Despite the efforts of the PNC to communicate information to the DEVs in the piconet via the beacon announcements, some of the DEVs may miss that information. Because the 802.15.3 piconet operates in a dynamic wireless medium, it is likely that these announcements will be missed from time to time and so it is important that a method is provided for the DEVs to get any information that they missed. Table 4–6 provides a list of the beacon announcement IEs and an alternate method for retrieving the information in that IE.

Table 4–6: Alternative methods to retrieve beacon announcement IEs

Information element	Alternative method to get information
DEV Association	PNC Information Request command
PNC Shutdown	Beacons disappear and ATP expires
Piconet Parameter Change	New information is contained in every beacon
Application Specific	Vendor specific
Pending Channel Time map (PCTM)	IE is in every beacon as long as it is needed
PNC Handover	New PNC DEV address is in the new beacons
CTA Status	Probe Request command
PS Status	IE is in every beacon as long as it is needed

Dynamic channel selection

The wireless environment of 802.15.3 is dynamic and uncontrolled. As an unlicensed user, 802.15.3 DEVs need to be able to handle interference from incumbent users in the frequency band of operation. However, if the PNC determines that the characteristics of the current PHY channel are poor, it can search for a new channel for the piconet. The PNC has three methods by which it can determine the characteristics of the available PHY channels:

a) The PNC can scan all of the available PHY channels (see "PNC channel scanning" on page 150) to determine the level of interference.

b) The PNC can use the Remote Scan Request command (see "Remote scan" on page 152) to ask another DEV in the piconet to scan some or all of the PHY channels.

c) The PNC can use the Channel Status Request command (see "Channel status" on page 148) to ask the DEVs in the piconet to return statistics on the current connections in the piconet.

Although the current PHY uses frequency-based channelization, this method is applicable to other types of channelization, including code-based channelizations like CDMA. The PHY is responsible for mapping the concept of a channel number into the precise characteristics of the PHY channel.

Regardless of how the PNC arrives at the decision to change the PHY channel, the PNC must perform a PNC channel scan of the new channel before it moves the piconet. This last scan is important because the PNC is the center of the piconet, and so the PNC must determine that the new channel is sufficiently clear before it moves the piconet. Once the PNC has decided to change the piconet's PHY channel, it places the Piconet Parameter Change IE in the beacon with the new channel number and the beacon number of the first beacon that will be sent on the new PHY channel.

In the last superframe on the old channel, i.e., the superframe where the beacon number is one less than the beacon number in the Piconet Parameter Change IE, the PNC and the other DEVs will have to cease operations prior to the end of the superframe to ensure that they have enough time to switch channels. The PHY specification incudes a parameter that indicates the maximum time allowed for a channel change. In the case of the 2.4 GHz PHY, the DEVs are allowed a maximum of 500 μs to switch channels. Thus, in the last superframe, the last 500 μs of CTAs will go unused as the DEVs switch to the new channel. If possible, the PNC should arrange the CTAs in the last superframe such that there are not any allocations during the time that the DEVs require to change channels.

If the parent PNC changes channels, the dependent PNCs will have to either change channels at the same time as the parent or cease operations as a piconet. If there is only one dependent piconet, it would be possible for that piconet to remain in the old channel, eventually becoming an independent piconet. However, there may be more than one dependent piconet that is a part of the parent piconet. In this case, it is possible that more than one dependent piconet could remain in the old channel and that they would begin to collide with each other if they remained in operation. To avoid this potential problem, IEEE Std 802.15.3 requires that dependent piconets that want to remain in the old channel first shutdown operations and then restart the piconet in the channel.

The frame exchange sequence for changing the PHY channel is illustrated in Figure 4–32.

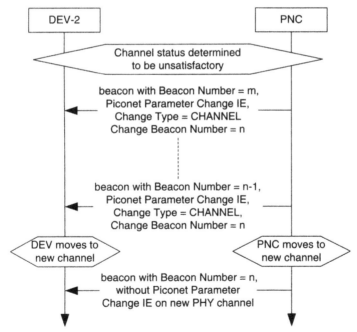

Figure 4–32:   Changing the PHY channel

Changing the PNID or BSID

DEVs in the piconet identify the beacon by checking three values, the PNID, the BSID, and the DEV Address of the PNC. Only when all three match the values that the DEV is looking for can it be certain that it has found the beacon for its piconet. In the case of the DEV Address, the values are required to be unique by the IEEE registration authority committee (RAC). A DEV in the piconet finds out about a change in this parameter via the PNC Handover IE that is sent during the PNC handover

> The IEEE RAC is a subcommittee of the IEEE Standards Association Standards Board (IEEE-SASB) and its purpose is to assign unique numbers to support standards. Among other things, the RAC is responsible for assigning MAC address groups to companies. The RAC assigns a 3-octet (24-bit) prefix number to companies or organizations to use as an organization-ally unique identifier (OUI). The company or organization then uniquely assigns the last 3 octets (for 48-bit MAC addresses) or the last 5 octets (for 64-bit MAC addresses) to products that they produce.

process. The BSID and PNID, on the other hand, may not be unique. In the case of the PNID, the PNC that starts the piconet is supposed to pick a random value for the PNID. However, this does not guarantee that the piconet will not encounter another piconet with the same PNID.

The problem of overlapping PNIDs is very important in 802.15.3 because the unique DEV address is not used in the addressing of frames. Thus, two piconets that are overlapping may have two different DEVs with the same DEVID. Each DEV checks not only for its DEVID as the DestID in the MAC header, but it is also required to ensure that the PNID matches the value that it expects. This prevents the two DEVs in different piconets from both trying to ACK the same frame or respond to the same command. Thus, the PNID is in each frame to help ensure the uniqueness of the 802.15.3 addressing scheme. However, this method does not work if the adjacent piconet also uses the same PNID. The first working group letter ballot pointed out this flaw, and the group developed techniques to enable the DEVs and the PNC to detect the presence of another piconet with the same PNID and to be able to change the PNID in an operating piconet.

If a DEV determines that there is another piconet within its operational area that is using the same PNID, that DEV is required to report that fact to the PNC. There are two ways that the DEV may determine that this is the case:

a) The DEV receives a beacon from a different PNC with the same PNID. The DEV will know that it is a different PNC by checking the PNC Address in the Piconet Synchronization Parameters field.

b) The DEV receives a frame with the current PNID where the SrcID or DestID is not a valid ID in the current piconet.

In the second case, the DEV needs to be careful that it does not report an overlapping piconet simply because it has not been able to keep up with the changes in the piconet membership.

If the DEV does detect an overlapping piconet with the same PNID as the DEV's current piconet, the DEV will use the Announce command to send the Overlapping PNID IE to the PNC to report this. The PNC will also be performing periodic scans of the current PHY channel as well as other PHY channels, and so it may detect an overlapping piconet with the same PNID as well. Finally, the Remote Scan Response command contains a list of all piconets that the DEV was able to find in its search of the PHY channels. The PNC can examine the Piconet Description Sets that are returned in this command to see if there are any other piconets nearby that are using the same PNID.

If the PNC determines, either by a report from a DEV or from its own scans, that there is an overlapping piconet in any of the valid PHY channels, it is required to change the PNID. Note that this requirement applies even to piconets that are currently occupying a PHY channel that is different from the current one. Although the overlapping piconet may currently be in a different PHY channel, it could change channels into the piconet's PHY channel. Therefore, the PNC must change the PNID to avoid the addressing problems that would occur if the overlapping piconet that is using the same PNID changed into the current channel.

Unless an overlapping piconet is found, the PNID remains constant for the life of the piconet. If the PNC shuts down the piconet without handing over control, it can reuse the PNID from the previous piconet. Otherwise, the PNC needs to randomly select a new PNID for the new piconet. When the PNC hands over responsibility to a new PNC-capable DEV with the handover process, both the PNIDs and BSIDs remain the same.

Although the PNID is used in every frame, the BSID appears only in the beacon. The BSID is a text field of 6 to 32 ASCII characters that is intended to provide a user-friendly method to identify a specific piconet. If a DEV has lost track of the piconet for some time, it can regain synchronization with the piconet using the BSID. The PNID or the DEV address of the PNC may have changed in the meantime, but the BSID will likely remain constant. Using a text string instead of a hexadecimal number makes it possible for applications to display the names of piconets that are discovered in a manner that can be understood by a typical user, e.g., "Bob Jones' piconet" instead of a MAC address of 00-09-6B-50-3C-7A.

Because the BSID is used for scanning and by the user to select a piconet, the PNC is not required to change the BSID if it detects another piconet nearby that is using the same BSID. In the case of a home network, the user may wish for all of the piconets to use the same BSID. The PNC is allowed to change the BSID using the same procedure that is used to change the PNID. However, the PNC is not allowed to change both the PNID and the BSID at the same time.

The PNC changes the PNID or BSID using the Piconet Parameter Change IE with the Change Type field set to either "PNID" or "BSID" in the beacon. In the superframe indicated by the Change Beacon Number field, all of the DEVs in the piconet, including the PNC, will start using the new PNID or BSID. The announcement of this IE follows the same rules as other beacon announcements, including alignment with system wake beacons, if required.

Moving the beacon or changing the superframe duration

The PNC needs to be able to control the timing in the superframe, and so it is allowed to change the superframe duration or location of a beacon relative to a previous beacon. The PNC may change the superframe duration, as illustrated in Figure 4–33, to handle more traffic or to change the latency for allocations in the superframe.

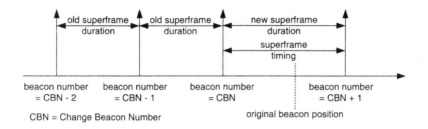

Figure 4–33: Changing the superframe duration

Likewise, the PNC may also wish to change the position of the superframe relative to previous superframes, as illustrated in Figure 4–34. This capability is required for dependent piconets because the parent PNC is allowed to move the CTA position for the dependent piconet. Thus, the dependent PNC needs to have a method by which it can inform DEVs in its piconet that the superframe will be changing position relative to a previous superframe. As the figure shows, the PNC is only allowed to delay the start of the next superframe, it cannot advance the start of a subsequent superframe. If the PNC were allowed to advance the start time of the following superframe, then the next superframe would overlap with the current superframe, potentially causing collisions with the last CTAs and the start of the next superframe. The PNC can achieve the same effect as advancing the superframe position by a small amount by setting the Superframe Timing value that is equal to the superframe duration minus the amount that the superframe position needs to move. Using only a positive value for the move also made the specification simpler in the draft.

The PNC notifies the DEVs in the piconet of a change in the superframe position or superframe duration by placing the Piconet Parameter Change IE in the beacon with the Change Type field set to either "MOVE" or "SIZE" and the Superframe Timing field set to the appropriate value. If the Change Type field is set to "MOVE," then the Superframe Timing field is the delay in the start of the superframe that has the beacon number indicated in the Change Beacon Number field. If the Change Type field is set to "SIZE," then the Superframe Timing field is the duration of the superframe that has the beacon number indicated in the Change Beacon Number field.

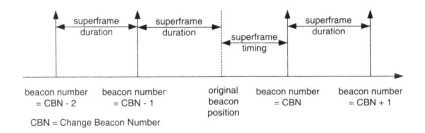

Figure 4–34: Changing the relative position of the superframe

A dependent PNC includes the Piconet Parameter Change IE in the beacon with the Change Type field set to "MOVE" and the Superframe Timing field set to zero when the parent PNC is handing over control of the piconet. During this procedure, the new PNC may move the dependent piconet's allocation when it takes over as the PNC, and the dependent PNC needs to inform its DEVs that a change is about to happen that may affect the timing in the piconet. The DEVs in the dependent piconet are all required to wait until they successfully receive a beacon from the dependent PNC following this change before they can transmit again. This rule is always in effect for DEVs with dynamic CTAs, so that this does not create a new requirement for those DEVs. However, DEVs in the dependent piconet with pseudo-static CTAs need to know that the location of their allocation may be changing as well.

FINDING INFORMATION

A key component of any protocol is that it provides a method for a participant to find out information about other participants in the network as well as the condition of the network. In 802.15.3, there are seven defined methods for sending or retrieving information about either the DEVs in the piconet or the current conditions in the piconet. These seven methods are as follows:

a) *Probe Request and Probe Response commands:* These commands are used to request IEs and to respond with the requested IEs.

b) *Announce command:* This command is used to send an IE to a DEV that did not request the IE.

c) *PNC Information Request and PNC Information commands:* These commands provide information about the capabilities of the DEVs that are members of the piconet.

d) *Channel Status Request and Channel Status Response commands:* These commands allow the DEVs to determine the link quality based on the number of lost frames.

e) *PNC channel scanning procedure:* The PNC searches all or a portion of the PHY channels to determine the best channel for the piconet.

f) *Remote Scan Request and Remote Scan Response command:* The DEVs in the piconet report the status of scanning a list of PHY channels.

g) *Piconet Services command and IE:* This command and IE provide support for service discovery. The DEVs report vendor-specific information about the higher layer capabilities that they provide.

Probe

The Probe Request and Probe Response commands are used to obtain an IE from the PNC or from another DEV in the piconet. In all but the last two drafts of the standard, a single command—the Probe command—took the place of the three commands now in the standard; Probe Request, Probe Response and Announce. This single command could be sent back and forth between two DEVs with no end in the frame exchange sequence, giving it a "ping pong" quality. This versatility made it almost impossible to accurately specify its behavior in the MLMEs and in the MSCs. The more that the task group worked with this "super command," the more that they disliked the way that it was defined. Finally, a proposal was presented that split the Probe command into the more manageable trio that is now used to request, respond and send IEs. A comparison of the description of the Probe command in previous versions of the draft with the descriptions of Probe Request, Probe Response, and Announce in the final version of the standard clearly show that the new commands are much simpler to describe and implement. This is most clearly demonstrated in the simplicity of the new MSCs versus the complex combination of MSCs in the older drafts which did not even completely describe the Probe command's behavior.

The procedure for requesting an IE from the PNC or another DEV in the piconet begins by sending a Probe Request command to a member of the piconet, as shown in Figure 4–35. This command contains an Information Requested field and a Request Index field. The Information Requested field is either a bitmap that identifies the IEs or a binary representation of an IE's element ID. When the Information Requested field contains a bitmap, the DEV is able to request multiple IEs as long as their element ID is less than 31[25]. Note that the Vendor Specific IEs all have element IDs larger than 0x80 and so they cannot be requested using the bitmap. Instead, these IEs must be requested one at a time with separate Probe Request commands.

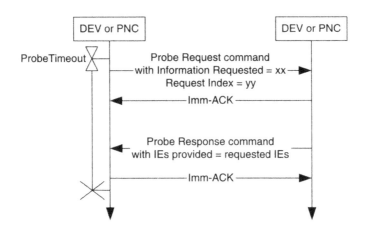

Figure 4–35:   DEV using Probe Request command to obtain an IE

The Request Index field is used in the Probe Request command to indicate the specific IE that is requested when more than one would be appropriate. This field was introduced to enable a DEV to request a specific CTA Status IE. The PNC is able to provide CTA Status IEs for each of the allocated streams in the piconet. Thus, when a DEV requests the

> The Request Index field is 2 octets although only a single octet is currently used and it has only been defined for a single purpose. The field was named generically and made larger so that it would support updates to the standard without changing the format of the command. In addition, the ASIE or vendor specific IEs may also use this field and it was felt that it was better to allow more room in case it was needed.

---

[25] There are currently only 16 IEs specified in the standard, but this may be increased in future revisions.

CTA Status IE from the PNC with a Probe Request command, the PNC needs to know if the DEV wants all of the possible CTA Status IEs or if it just wants information about a particular Stream Index. The Request Index field in this case contains the Stream Index for which the DEV is requesting the CTA Status IE. If least significant octet of the Request Index field is the BcstID, then the DEV is requesting that the PNC send all valid CTA Status IEs to the DEV.

When a DEV receives a Probe Request command, it first checks to see if it is allowed to respond to the request. The DEV is forbidden to respond to a request for some IEs. If the DEV is allowed to respond with any of the IEs that were requested, it sends a Probe Response command to the requesting DEV with the some or all of the requested IEs. The DEV is allowed to split its response over multiple Probe Response commands if the length of the IEs exceeds the maximum frame length or if the DEV does not want to send long Probe Response commands due to the channel characteristics. However, the Probe Response command cannot be fragmented.

Some of the IEs used in the standard are not intended to be sent or received from DEV to DEV or they may have validity for only a limited period of time, e.g., the CTA IE, Piconet Parameter Change IE or the PS Status IE. Therefore, the DEVs are forbidden to request certain IEs from other DEVs or from the PNC, depending on the type of IE and whether the entity making the request is a regular DEV or is the PNC. The rules for requesting IEs are summarized in Table 4–7.

> **Notice** When the now obsolete Probe command was being used, a DEV requested a CTA Status IE by first sending a CTA Status Request IE to the PNC with the stream index. The PNC then responded with the correct CTA Status IE(s). Because of this, it was not possible to directly request either the CTA Status IE or the CTA Status Request IE. Unfortunately, when the standard was updated to use the new Probe Request/Probe Response method, the tables that indicate the allowed requests and responses was not updated. Thus, there is an error in these tables in the standard; a DEV is allowed to request this IE from the PNC and the PNC is allowed to respond. The restrictions are correctly listed in these tables.

Similarly, not every request that a DEV receives in a probe command should receive a response. In this case, the rules are different depending on the source and destination of the request. For example, it makes sense for the PNC to request the Overlapping PNID IE from a DEV, but a DEV should not request it from the PNC.

When either a DEV or the PNC receives an IE request in a Probe Request command, it is either required to respond to the request, allowed to respond to the request, or forbidden to respond to the request. The rules for responding to a request for an IE in the Probe Request command are summarized in Table 4–8. Note that there are only three types of requests possible. The fourth, PNC requesting from the PNC, is not a possible interaction.

Table 4–7: Rules for requesting IEs in a Probe Request command

Information element	PNC allowed to request?	DEV allowed to request?
CTA	Shall not request	Shall not request
BSID	Shall not request	May request
Parent Piconet	Shall not request	May request
DEV Association	Shall not request	Shall not request
PNC Shutdown	Shall not request	Shall not request
Piconet Parameter Change	Shall not request	Shall not request
Application Specific	May request	May request
PCTM	Shall not request	May request
PNC Handover	Shall not request	Shall not request
CTA Status	Shall not request	May request
Capability	May request	May request
Transmit Power Parameters	May request	May request
PS Status	Shall not request	Shall not request
CWB	Shall not request	Shall not request
Overlapping PNID	May request	Shall not request
Piconet Services	May request	May request
Vendor Specific or Reserved	May request	May request

Table 4–8: Rules for responding to requests for IEs in a Probe Request command

Information element	DEV received request from DEV	DEV received request from PNC	PNC received request from DEV
CTA	Shall ignore	Shall ignore	Shall ignore
BSID	Shall ignore	Shall ignore	Shall respond
Parent Piconet	Shall ignore	Shall ignore	Shall respond
DEV Association	Shall ignore	Shall ignore	Shall ignore
PNC Shutdown	Shall ignore	Shall ignore	Shall ignore
Piconet Parameter Change	Shall ignore	Shall ignore	Shall ignore
Application Specific	May respond	May respond	May respond
PCTM	Shall ignore	Shall ignore	Shall respond
PNC Handover	Shall ignore	Shall ignore	Shall ignore
CTA Status	Shall ignore	Shall ignore	Shall respond
Capability Information	Shall respond	Shall respond	Shall respond
Transmit Power Parameters	Shall respond	Shall respond	Shall respond
PS Status	Shall ignore	Shall ignore	Shall ignore
CWB	Shall ignore	Shall ignore	Shall ignore
Overlapping PNID	Shall ignore	May respond	Shall ignore
Piconet Services	May respond	May respond	May respond
Vendor Specific or Reserved	May respond	May respond	May respond

Announce

The Announce command is used to send unsolicited information to another member of the piconet. Because the information was not requested, the sending DEV does not receive a response that the information was received, other than the Imm-ACK. The process of sending the Announce command is illustrated in Figure 4–36. The destination DEV is allowed to drop the incoming Announce command for various reasons, e.g., its internal buffers are full, it is too busy, etc., even if it ACKed the frame. Thus, the sending DEV may need to use another method, not explicitly specified in the standard, to verify that the IE was received and processed. The only IE where this is really important is for the Overlapping PNID IE. In this case, the DEV will know that the PNC received this IE correctly when it receives a Piconet Parameter Change IE in the beacon that is changing the PNID.

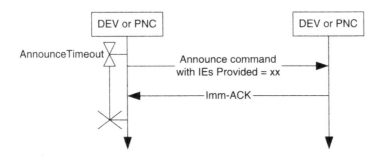

Figure 4–36: DEV using Announce command to send information

Not all of the IEs are allowed to be sent unrequested to another DEV in the piconet with the Announce command. For example, the CTA IE is sent in every beacon, but it is not allowed to be sent in the Announce command. The CTA IE is meaningful only at the beginning of the superframe when it is sent in the beacon. If it is sent in an Announce command during the superframe, it would be possible that the time in the CTA IE had already passed. Even if the allocated time in the CTA was after the reception of the Announce command, the DEV might not have enough time to prepare to use the CTA. For other IEs, it only makes sense if either the PNC (e.g., PS Status IE) or the DEV (e.g., the Overlapping PNID IE) sends it. The rules for sending IEs to other DEVs in the piconet are listed in Table 4–9.

Another use of the Announce command is to send it to a DEV to determine if it is still in the piconet. To do this, the DEV sends an Announce command with an empty IE provided field to another DEV in the piconet. The Imm-ACK response by that DEV verifies that it is still in the piconet and is within the range of the requesting DEV. If the target DEV is in a power-save mode, the originator will need to send the command in one of its wake beacons.

Table 4–9: Rules for sending unrequested IEs in an Announce command

Information element	PNC allowed to send?	DEV allowed to send?
CTA	Shall not send	Shall not send
BSID	Shall not send	Shall not send
Parent Piconet	Shall not send	Shall not send
DEV Association	May send	Shall not send
PNC Shutdown	May send	Shall not send
Piconet Parameter Change	May send	Shall not send
Application Specific	May send	May send
PCTM	May send	Shall not send
PNC Handover	May send	Shall not send
CTA Status	May send	Shall not send
Capability Information	May send	May send
Transmit Power Parameters	May send	May send
PS Status	May send	Shall not send
CWB	Shall not send	Shall not send
Overlapping PNID	Shall not send	May send
Piconet Services	May send	May send
Vendor-Specific or Reserved	May send	May send

PNC Information

When a DEV associates with a PNC, it passes a set of information that identifies it and indicates its capabilities as both a DEV and as a PNC, if it is PNC capable. The PNC needs to distribute this

In a non-secure piconet, a DEV becomes a member when it completes the association process. In a secure piconet, however, a DEV becomes a member once it has received a management key from the PNC. The DEV gets this key using a process that is outside of the scope of the standard. The other DEVs in the piconet will be informed that a new DEV is a member of the piconet when they receive a PNC Information command from the PNC with the new DEV listed as a member.

information to the piconet and needs to tell the new DEV about the capabilities and identities of the other DEVs in the piconet. To accomplish this, the PNC uses the PNC Information command to broadcast the information about the DEVs in the piconet whenever a new DEV becomes a member of the piconet. The broadcast PNC Information command contains the DEVID, the DEV address, the Overall Capabilities field, the DEV's ATP, and the DEV's requested System Wake Beacon Interval for each of the DEVs in the piconet. In addition, each entry for a DEV indicates if the DEV is a member of the piconet or if it is only associated with the piconet. The PNC also uses this command to send information about the DEVs in the piconet to the new PNC in the PNC handover process (see "Handing over control" on page 66).

For the other DEVs in the piconet, the PNC Information command carries some key information. The most important information is the pairing of the DEVID and DEV address. Because a DEV might join the piconet multiple times, it could end up using different DEVIDs. The other DEVs in the piconet will use the DEV address associated with the DEVID to determine the DEVs with which they need to communicate. Similarly, the higher-layer protocols will identify the DEVs by the unique MAC address, as the DEVID is only used in the MAC layer.

There are also two PHY-related characteristics that are contained in the overall capabilities field that is a part of a DEV Info field in the PNC Information command. The first characteristic is the data rates that are supported by the DEV. The PHY may have one or more mandatory data rates, but it may also have some optional data rates. If the DEV needs to use one of the optional data rates, then it must first determine that the destination supports this data rate using the information in the DEV Capabilities field. The second characteristic is the preferred fragment size. This field indicates to the other DEVs in the piconet the fragment size that this DEV prefers for incoming data frames.

The Overall Capabilities field included for each DEV in the PNC Information command also contains three pieces of information that can be used to help schedule a DEV's transmissions. When a DEV associates, it has the option to indicate that when it is in ACTIVE mode, it will be listening for frames addressed to it during certain portions of the superframe. Three options are available in the Overall Capabilities field:

- *Always AWAKE*—The DEV will be listening during the entire superframe for frames addressed to it.

- *Listen to Source*—The DEV will listen during all CTAs where the SrcID is the DEVID of a DEV which it is already the target of a CTA. For example, if DEV-1 sets the Listen to Source bit to be true and there is a CTA assigned with DEV-2 as the source and DEV-1 as the destination, DEV-1 will listen to all CTAs with DEV-2's DEVID as the SrcID, even if DEV-1 is not the destination.

- *Listen to Multicast*—The DEV will be listening to all Multicast CTAs, regardless of the SrcID or Stream Index.

The purpose of these three indications is that they make it possible for the source DEV to use all of its collective channel time to handle variable transmission demands to various destinations. For example, some video encodings result in a variable bit rate stream. If a DEV was sourcing three different video streams, it could request CTAs that were only large enough for the average bit rate of each of the three streams. If the destinations were listening to all of the source DEV's allocations, the source would be able balance the load among the three video streams, using left over time from a stream that was currently operating at a low data-rate to handle extra traffic from a stream that was currently operating at a high data-rate. The standard does not specify how a DEV might schedule the data for this, but by providing a method by which DEVs can indicate when they will listen, an implementation is given the opportunity to take advantage of this to improve throughput. One consequence of stating that the DEV will be listening more often than it needs to is that the DEV will use more power because its receiver will be on more often. The implementer will need to weigh the through-put advantages against the increase in power usage based on the target application and usage of the DEV.

Because a broadcast command is not ACKed, the PNC cannot guarantee that every DEV in the piconet will be able to receive the PNC Information command. If a DEV needs the information about either a single DEV or about all of the DEVs in the piconet, it sends the PNC Information Request command to the PNC. If the requesting DEV needs the information for a single DEV, it sends the request with the Queried DEVID field set to the DEVID about which it is seeking information. A DEV might send this request if it sees a CTA with a new DEVID or receives a frame from a DEVID that it does not recognize. When the DEV determines that there is a DEVID for which it does not have the appropriate information, it will send the request to the PNC to get this information. If the DEV has determined it is out of sync with the membership list of the piconet, it sets the Queried DEVID field to the BcstID. The PNC will recognize this as a request for the entire list of associated DEVs, including the fields corresponding to the PNCID and the NbrIDs, if any neighbor PNCs are associated with the piconet. A new DEV that joins the piconet but misses the broadcast of this command would also request the entire list so that it will know the membership of the piconet as well as the relevant DEV capabilities. The frame exchange involved in requesting and receiving information from the PNC using the PNC Information Request and PNC Information command is illustrated in Figure 4–37.

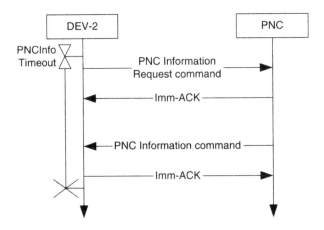

Figure 4–37: Requesting and receiving DEV related information from the PNC using the PNC Information Request and PNC Information commands

Channel status

In order to maintain the QoS requirements for the higher layers, the MAC and PHY need to be able to determine the quality of the link between two DEVs. The link quality affects the QoS in two ways. First, the source DEV needs to select a data rate that will have the highest throughput. If using a higher data rate increases the FER, then the DEV may have to use a lower PHY data rate to improve the

FER. Second, the DEV will use the data rate information along with the FER and the characteristics of the data to calculate the amount of time required in each superframe to support the data throughput required by the application.

The DEV can determine some information about the FER and, hence, the link quality by keeping track of the number of retransmissions required for frames that it sends to the destination. That gives part of the picture, but it does not include information about the number of times that the destination DEV ACKed the frame but where the source DEV was unable to hear the ACK. The Channel Status Request command provides a method for the originator DEV to request link quality information from the target DEV. The target DEV responds to the originator with the statistics of the number of frames it attempted to transmit, the number of frames it received, the number of those that were in error, and the number of no-ACK frames that were missed. The target DEV also indicates in the Measurement Window Size field the length of time over which this information was gathered.

In the totals reported in the Channel Status Response command, the target DEV will only count directed frames that were sent between the two DEVs. The DEV will count a frame as having been received in error if the HCS validation passes but the FCS validation fails. The DEV will keep track of missing no-ACK frames using the gaps in the MSDU numbers and report this number in the RX Frame Loss Count field. Because asynchronous traffic shares the same MSDU numbering among multiple destinations, the RX Frame Loss Count field only applies to isochronous streams.

The frame exchange sequence for determining the channel status between two DEVs is illustrated in Figure 4–38.

The PNC can also use the Channel Status Request command to collect information from all of the DEVs in the piconet. The PNC would use this information to help determine the quality of the current channel. To do this, the PNC sends the Channel Status Request command as a broadcast frame, i.e., with DestID set to the BcstID and the ACK Policy field set to no-ACK. DEVs that correctly receive this command will respond to the PNC with a directed Channel Status Response command when they get an opportunity to send commands to the PNC. The DEVs will report the aggregate status for all streams that have that DEV as the destination. If the PNC is not using the CAP for commands, then it will need to allocate MCTAs to each of the DEVs in the piconet to allow them time to respond to the broadcast request. Not all of the DEVs in the piconet will receive this command, and because it is sent with no-ACK policy, the PNC will not know which DEVs were unable to receive the command until it begins to receive the responses. In particular, DEVs in a power-save mode will likely not receive the message. However, because these DEVs are saving power, it is also likely that their traffic is light and the statistics would not be as meaningful as those from ACTIVE mode DEVs.

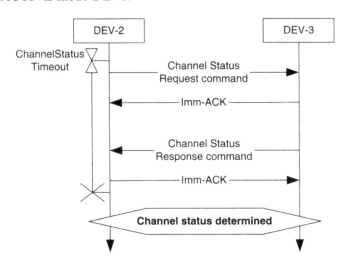

Figure 4–38:   Requesting channel status information from another DEV

PNC channel scanning

In addition to maintaining the timing and granting channel time to DEVs, the PNC also needs to make sure that the piconet is operating in the best possible channel. Only the PNC can make the decision to change the channel, and so the PNC needs to have as much information as possible about not only the current channel, but also about other channels that it could potentially use. Every PNC-capable DEV is required to be able to scan the PHY channels to determine the ones that already have piconets before it can start its own piconet. The PNC channel scanning procedure builds on this ability to give the PNC channel status information while the piconet is in operation.

The PNC, or any DEV for that matter, is always allowed to use time when it is not required to be receiving or transmitting to passively scan any PHY channel. The PNC should periodically scan the current channel during quiet times in the superframe to detect overlapping piconets or other sources of interference. The PNC can create "quiet" time in the superframe either by placing a specific allocation in the beacon with the SrcID and DestIDs set to the PNCID or by leaving gaps in between CTAs in the superframe. However, if the PNC wants to scan other channels with this procedure, it needs to stop sending the beacon for one or more superframes. The PNC stops the traffic in the current piconet because the energy from transmissions in the current channel in general will be nonzero in other PHY channels. If the piconet continues operation, the PNC may consider a DEV that is in close proximity to the PNC as interference from another piconet. The PNC could potentially not select that channel even though it was empty.

Although most of the DEVs in the piconet will cease transmitting when the beacon stops, DEVs with pseudo-static allocations will continue operation. The PNC can handle this issue in one of three ways:

a) The PNC can place the Piconet Parameter Change IE in the beacon with the Operation Type Set to "MOVE" and the Superframe Duration field set to zero. This will cause DEVs with pseudo-static allocations to stop transmitting after the Change Countdown field reaches zero. The DEVs will restart transmissions when the PNC begins sending beacons again.

b) The PNC can keep track of the times allocated as pseudo-static CTAs and ignore interference data collected during these times.

c) The PNC can stop sending beacons for at least mMaxLostBeacons before collecting the channel status information.

Once the PNC has scanned the alternative channels, it returns to the current channel and begins sending the beacon again at the time when a beacon would have been expected. The PNC will increment the beacon number so that it takes on the value that it would have been if the PNC had not stopped sending the beacon. Having completed the PNC channel scanning procedure, the PNC has four options:

a) Remain in the current channel.

b) Change the piconet channel (see "Dynamic channel selection" on page 132) to a channel that the PNC scanned.

c) Change the maximum TX power allowed in the piconet (see "Transmit power control" on page 157).

d) Perform another, unspecified, action.

The final option was provided so that the standard would not unnecessarily restrict implementers in developing new techniques for handling interference in the PHY channel. The algorithm for determining when the PNC will change the channel is outside of the scope of the standard.

The PNC is required to perform the PNC channel scanning procedure before changing the channels because it is the center of the piconet. The PNC may find through the Remote Scan Request/Remote Scan Response commands (see "Remote scan" below) that another channel is best for most of the DEVs in the piconet. However, if that channel is not also a good one for the PNC, it will not matter that it is a good channel for the other DEVs in the piconet. The PNC must be able to send and receive commands reliably in order to act as the coordinator of the piconet. Thus, in the end, the PNC must confirm with its own scanning that the new channel is better for piconet operations than the current one.

Remote scan

Although it is of paramount importance that the PHY channel for the piconet is relatively quiet from the PNC's point of view, it is also important for the PNC to be able to determine the characteristics of the channel at points in the piconet other than where it is. There may be two channels where the location near the PNC is relatively quiet, but one channel may have an interferer that is relatively close to DEVs that are near the outside edge of the piconet's operational area. The PNC may not be able to detect this interference even though it is causing problems with traffic for those DEVs. Using another DEV to perform the scan also allows the PNC to continue the piconet operations uninterrupted while still receiving information about potential sources of interference in the current channel and the status of other PHY channels.

The PNC uses the Remote Scan Request command to ask a DEV to perform a scan of one or more PHY channels. The Remote Scan Request command contains a list of the channels that the PNC is requesting that the DEV scan. The PNC may also place quiet time in the beacon, as described in "PNC channel scanning" on page 150, for the DEV to perform the channel scan. If the DEV accepts the request, it uses the normal channel scan procedure that it would use to find a piconet to join. The DEV will keep track of all of the piconets that it finds and records pertinent information about them. The DEV will pass back the PNID of the piconet that it found and the channel index where it found the information. If the DEV found only a frame from the other piconet, this is the only information that it can report. If the DEV receives a beacon from another piconet, it will also report the DEV address of the PNC, the BSID of the piconet, and the Parent Piconet IE, if it is present in the beacon. The DEV will indicate if it found either a beacon or non-beacon frame from another piconet and if the piconet is a dependent piconet.

In addition to reporting all of the piconets that the DEV found, the responding DEV will also rate the channels that it scanned for interference, listing them from best (least interference) to worst (most interference). The DEV does not respond with an absolute measure of the interference nor does it give a relative rating or interference, i.e., that channel 4 is three times as bad as channel 2. An absolute measurement of interference is difficult to define. The DEV would need to identify the type of interference it found; i.e., is it wideband or narrowband? Then it would need to report the percentage of time that the interference appeared. Finally, the PHY would need to be calibrated so that it gave relatively accurate measurements for the interference. Instead of creating all of these requirements, which would make it difficult to develop low-cost implementations, the standard requires only that the DEVs give an ordered rating using whatever criteria are appropriate for that DEV.

When the DEV has completed the scan, it will respond to the PNC with the Remote Scan Response command containing a listing of the channels that it scanned, the relative ranking of the channels for interference, and the list of all of the piconets that it found. The PNC is then able to use this information in determining if a channel change is required and to help in selecting the new channel if it needs to change channels.

Because channel scanning can take a long period of time, the DEVs are allowed to refuse the request to scan the channels for the PNC. If the DEV is not going to perform the channel scan, it responds to the PNC with the Remote Scan Response command with a Reason Code field set to "Request denied." If for some reason the PNC requested a scan of a channel that is not valid, the DEV will respond with the Remote Scan Response command with the Reason Code field set to "Invalid channel requested." The frame exchange sequence for the PNC to request a remote channel scan and for the DEV to respond with the information is illustrated in Figure 4–39.

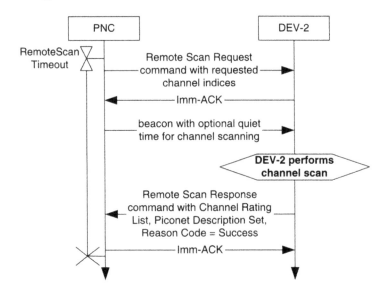

Figure 4–39:   PNC requesting that a DEV perform a remote scan

A DEV may also send an unsolicited Remote Scan Response command to the PNC if it is receiving a beacon from another piconet. The DEV will not report any piconet that it determines to be a child or neighbor of the current piconet. It is possible, however, that the DEV will mistakenly report a child or neighbor piconet as an overlapping piconet. It is up to the PNC to determine which reports that it receives from the DEVs in the piconet are from overlapping piconets as well as the action required to handle any interference issues.

Piconet services

From a user's point of view, the most important characteristic of a DEV in the piconet is the application level services that the DEV provides. For example, when a video camera finds a piconet, it will want to know if any of the DEVs have nonvolatile storage capabilities, e.g., a hard-drive, DVD writer, etc., or display capabilities, e.g., a flat-panel display or television. The ability of one entity to determine the application level characteristics of another entity is referred to as service discovery, and the process to determine this information is a service discovery protocol. Because application level information is outside the scope of a MAC/PHY standard, IEEE Std 802.15.3 cannot provide a complete specification of a service discovery protocol. There are higher layer protocols that do support service discovery, e.g., IEEE Std 1394 [B17], [B18], Jini™, or Universal Plug and Play™, so it is not necessary for 802.15.3 to provide another incompatible definition.

However, it is important for DEVs to be able to determine very early on the capabilities that are offered in a piconet. Ideally, the DEV would be able to determine these by passive scanning, but continually broadcasting these capabilities would needlessly waste bandwidth. Instead, IEEE Std 802.15.3 provides a method in the association process whereby an associating DEV can request the broadcast of the collected capabilities in the piconet. The goal is that within a very short period of time, ideally less than 1 second, a DEV can find a piconet, join it, and determine the application level services that are available. As an example, a person carrying a digital camera should be able to walk in the vicinity of a photo finishing kiosk and, within 1 second, find an icon for that kiosk on the camera's display. The user could then select print, select the kiosk, and the pictures would be transferred quickly to the kiosk for printing. The ability to quickly provide useful information to the user has always been a key criteria in the development of IEEE Std 802.15.3.

If an associating DEV requests the piconet services information in the Association Request command, the PNC will broadcast the Piconet Services command with all of the Piconet Services IEs that the PNC has collected from DEVs in the piconet. Because IEEE Std 802.15.3 does not define the format of this data, each IE includes an organizationally unique identifier (OUI) to enable the DEV that receives this command to determine if it can understand the contents and to determine how to process the contents if the DEV does support the OUI. If the PNC supports the Piconet Services command, it only has to store the IEs. The PNC is not required to be able to understand any of IEs that it stores and it can set a limit on the number or total size of the information that it will store.

Even though the PNC may not understand the contents of the IEs, this still provides a very useful service to the DEVs in the piconet. For example, suppose a DEV made by Manufacturer A joins a piconet with a PNC made by Manufacturer B that supports the Piconet Services command but does not support the decoding of Piconet Services IEs from Manufacturer A. When the DEV associates and requests the list of the piconet services, the command might contain three IEs from DEVs made by Manufacturer A that list services that the DEV needs. In this case, the DEV would be able to quickly, i.e., in less than 1 second, find out that a piconet offered an important service that it needed, e.g., printing, storage, WAN connectivity, and so on.

The capabilities of a DEV may be a security concern. For example, if the PNC in a home network always responded to a request for the Piconet Services by listing the number of TVs, VCRs, set-top boxes, and so on, it would be providing an inventory to potential thieves. On the other hand, some DEVs will always want to broadcast their services because the owners use them for making money. Examples of these types of entities would include photo-printing kiosks, public internet access hot-spots, and so on. The standard allows the DEV to refuse to broadcast this information either as a DEV or as the PNC. This allows the user to control the level of openness provided by these commands and IEs.

If a DEV wants to add its services to the list for the piconet, the DEV sends the Piconet Services IE to the PNC in an Announce command. If the PNC supports storing the Piconet Services IEs and it has storage space for the IE, it will save the IE for when the next DEV requests the information. The PNC will also send the Piconet Services IE to all the DEVs in a broadcast Announce command so that the existing DEVs in the piconet will see the services offered by the new DEV.

If a DEV sends a Probe Request command to the PNC requesting the Piconet Services IE, the PNC will respond to the DEV with Probe Response commands that contain the Piconet Services IEs for all of the DEVs in the piconet. If a DEV did not send a Piconet Services IE to the PNC, the PNC will place a Piconet Services IE in the Probe Response command with the DEV's DEVID and a zero Vendor OUI field and zero length Piconet Services field. If the PNC did not have sufficient space to store the Piconet Services IE submitted by the DEV, it will put a Piconet Services IE in the Probe Response command with only the DEV's DEVID; i.e., the IE will have length 1. Thus, the requesting DEV will receive an

IE for every DEV in the piconet, and it will know if any are missing in the response as well as knowing when the response is complete.

A DEV may also request a Piconet Services IE from another DEV in the piconet using the Probe Request command. If the DEV supports the Piconet Services IE and its current policy allows the broadcast of this information, the DEV will respond by sending its Piconet Services IE in the Probe Response command. Otherwise, the target DEV will respond with Probe Response command that contains a zero length Piconet Services IE which is an IE with only the Element ID and Length fields.

OTHER CAPABILITIES

IEEE Std 802.15.3 also provides a variety of capabilities that enhance the performance of the standard. This section describes how the PNC and DEVs can control the transmit power in the piconet as well as the rules for handling DEVs that support more than one data rate.

Transmit power control

IEEE Std 802.15.3 provides two types of transmit power control: a piconet-wide limit on transmit power and a negotiated level for DEV-to-DEV communications. The transmitter power used in the beacon and the CAP help to determine the physical size of the piconet. In order to constrain the size of the piconet, the PNC may set a maximum transmitter power that applies to the beacon, the CAP, and any directed MCTAs. The PNC is allowed to set power level to any value as long as it is not less than pMinTPCLevel, which is a PHY-dependent parameter. For the 2.4 GHz PHY, this parameter has a value of 0 dBm, which the lowest TX power level that a compliant DEV is required to support for this PHY. The PNC communicates the maximum TX power level to the DEVs in the Max TX Power Level field in the beacon. The DEVs have up to the 10*mMaxLostBeacons (= 40) superframes after receiving a beacon with a change to this field to comply with the new power level. The DEV is required to set its TX power to a level less than the value in this field. If the PNC is not going to put a limit on the maximum transmit power, it sets this field to its maximum value of 0x7F,[26] which corresponds to

---

[26] The field has units of dBm and is encoded in two's complement notation, so it has a minimum value of 0x80 or –128 dBm and a maximum value of 0xFE or +127 dBm.

+127 dBm (which will likely exceed the realistic transmit power of any imaginable PHY).

In addition to the piconet-wide limit on transmit power, DEVs that are exchanging data in a CTA can also negotiate a change in the transmit power level. Because this occurs in a specific CTA, a change in the transmit power will not adversely affect the other DEVs. This is not true during a CSMA/CA period like the 802.15.3 CAP or in an 802.11 WPAN. During a CSMA/CA period, all of the participants need to transmit with as much power as possible. If one of the participants was to decrease its transmit power, then other DEVs in the piconet would not sense the frame transmission and therefore would not suspend the backoff counter, leading to more collisions. Because a CTA is a TDMA allocation, the DEVs using the allocation can negotiate a lower transmit power to reduce interference and save power without adversely affecting the network or their throughput.

A DEV requests a decrease or increase in another DEVs transmit power by sending the Transmit Power Change command to the other DEV. This command includes a field that specifies a positive or negative change in decibels to the other DEV's transmit power. The field is encoded in two's complement notation, which allows a range of requested changes from –128 dB to +127 dB[27]. Because the requesting DEV would not know the current transmit power of the remote DEV, it simply asks for a change from whatever the current setting is. The actual transmit power that the other DEV is using is not relevant to the receiver; only the level of the received power relative to the receiving DEV's sensitivity is important. Thus, sending requested changes to the remote DEV is a simple and effective way to communicate a request for a transmit power change.

When a DEV receives a request for a transmit power change, the DEV will set its power level as close as possible to the requested change. The DEV will use this new power level for all CTAs that are assigned to the two DEVs. The message exchange for negotiating a different power level between two DEVs is illustrated in Figure 4–40.

---

[27] This range should be sufficient for any realistic PHY.

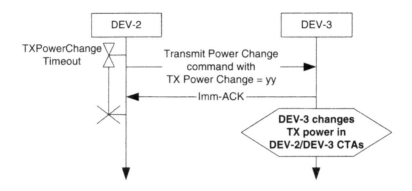

Figure 4–40: Requesting a change in a DEV's transmit power

Multirate capabilities

IEEE Std 802.15.3 was developed to support PHYs with multiple data rates where some of the data rates are optional. In the case of the 2.4 GHz PHY, there is one mandatory or base rate, 22 Mb/s, and four optional rates, 11, 33, 44, and 55 Mb/s. Because not all DEVs support the optional rates, broadcast and any traffic where the supported data rates of the DEV are unknown have to be sent at the base rate. For example, the Association Request and Association Response commands must be sent at the base rate because the DEV may not know the rates that the PNC supports and the PNC may not know the rates that the associating DEV supports. Once a DEV has joined the piconet, it can find out the supported data rates of another DEV in the piconet by using any of the following methods:

a) Check the DEV capabilities field for the DEV in the broadcast PNC Information command.

b) Send a Probe Request command to the DEV (at the base rate) requesting that the DEV send the Capability IE.

c) Send the PNC Information Request command to the PNC to request the record of the DEV.

Once the DEV has determined the data rates supported by the other DEV, it can use any one of the supported rates in directed communications with that DEV. The PNC's supported rates can be obtained either by examining the DEV Info field for the PNCID or by examining the DEV Info field that corresponds to the PNC's DEV personality by searching for the PNC's DEV address.

All ACKs are sent at the same rate as the frame that requested the ACK. In the case of the 2.4 GHz PHY, this only changes the ACK for an 11 Mb/s frame. In all other cases, the ACK is the same for all data rates because it only includes the MAC header, which is sent with the 22 Mb/s mode. In the 11 Mb/s mode, however, the MAC header field is sent twice, once at the base rate of 22 Mb/s and once at the 11 Mb/s rate. If the source DEV is using Dly-ACK, all of the frames in the burst must use the same data rate and the Dly-ACK frame is also sent at that rate. The rules for using a data rate are summarized in Table 4–10.

Table 4–10: Allowed PHY data rates for each frame type

Type of frame	Allowed PHY rates
All broadcast and multicast addressed frames (including the beacon and commands)	Base rate
Imm-ACK	Same rate as the frame that is being ACKed
Dly-ACK	Same rate as the last frame of the burst being ACKed
Association Request command	Base rate
Association Response command	Base rate
Disassociation Request command	Base rate
Directed command Frame	Any rate supported by both the source and destination
Directed Data Frame	Any rate supported by both the source and destination

EXTENSIBILITY OF THE STANDARD

IEEE Std 802.15.3 was developed with the understanding that it would need to be both flexible and extensible. To accomplish this goal, the standard provides specific IEs and commands that are reserved for vendor-specific extensions that are either outside of the scope of the standard or are yet to be defined. In the case of 802.11, the standard reserved all of the unused IE numbers for future use in the standard. However, the vendors needed to extend the standard with additional functionality, and so they began to use some of the reserved IE numbers. As IEEE Std 802.11 began to use more of the numbers to support enhancements to the standard, it increased the possibility of using a number already in use by a specific vendor.

IEEE Std 802.15.3 has explicitly called out a set of IE element IDs and command types that are reserved for vendor-specific use. In addition, IEEE Std 802.15.3 requires that the vendors identify these elements in standardized manner using a unique identifier to prevent interoperability problems with DEVs made by different manufacturers. Each Vendor Specific command and Vendor Specific IE, as well as the application specific information element (ASIE), all have an OUI field following the length field of either the command or IE. The OUI is defined to be the organizationally unique identifier issued by the IEEE RAC. Every manufacturer will be required to get an OUI in order to issue DEV addresses for the products that they create. They can use the same OUI to provide enhanced functionality for DEVs that understand these IEs or commands.

The inclusion of the OUI is necessary to prevent interoperability problems with other DEVs that are also using the Vendor Specific commands. Without it, a DEV might think that a Vendor Specific command that is supposed to say "respond if you are a printer" instead says "begin transmitting continuously for test mode." By using an OUI, the vendors will ensure that there is no confusion in the commands and IEs. If not, that vendor will experience a high return rate of their products, which will quickly end their participation in the market.

The first element that was proposed to provide extensibility to the standard was the ASIE. The ASIE was introduced to enable an implementation to place information in the beacon to be used by DEVs that supported the features offered by the PNC. The PNC can put multiple ASIEs in the beacon, and they can be placed either in the beacon frame or in the beacon extension. The contents of the ASIE are specific to the OUI and are outside the scope of the standard.

The standard has also set aside a block of element IDs for IEs and command types for commands. For IEs, the element IDs of 0x80–0xFF are reserved for vendor-specific IEs and will never be used by the standard (the 122 still available for the standard should be sufficient). In the case of vendor-specific commands, a modest 65,000+ command types are available for use as vendor-specific commands. The command type is two octets, and the standard has only used or reserved for use the first 256. Again, this should be sufficient for the foreseeable future for 802.15.3. If the standard ever did run out of element IDs or command types, it could overcome this problem by securing an OUI and using this to open up new element IDs or command types.

The content and use of the vendor-specific IEs and vendor-specific commands is outside the scope of the standard.

EXAMPLE OF THE LIFE CYCLE OF A DEV

The standard provides information on each of the various events that a DEV might encounter but does not put this together to provide a beginning to end description of the life cycle of a DEV. Because the application requirements for each DEV are different, the life cycle of each DEV will be different as well. This section provides a set of MSCs that give an example of a typical DEV's interaction with a piconet. In the MSCs, the ACKs have been left out for clarity. The first MSC, Figure 4–41, shows a DEV that joins a piconet, allocates channel time and exchanges data with another DEV in the piconet.

The second MSC, Figure 4–42, illustrates the same DEV leaving the piconet and the actions that the PNC takes when a DEV leaves a secure piconet.

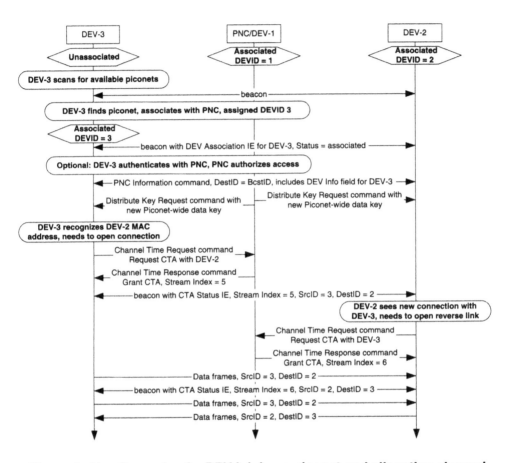

Figure 4–41: Example of a DEV joining a piconet and allocating channel time

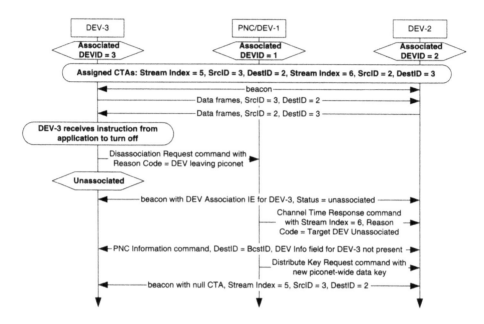

Figure 4–42:   Example of a DEV leaving a secure piconet

Chapter 5 Dependent piconets

INTRODUCTION

Dependent piconets were added to the standard to enable a variety of applications for IEEE Std 802.15.3. There are two types of dependent piconets defined in the standard: child and neighbor. The key difference between a child and a neighbor piconet is the membership status of the dependent PNC. A child PNC is a full member of the parent piconet. It can exchange data with other DEVs in the piconet and can send commands to any of the other DEVs in a piconet. A neighbor PNC, on the other hand, is not a full member of the parent piconet and so it is restricted in its communications in the parent piconet. The neighbor PNC is associated in the parent piconet, but it is not required to become a secure member of the parent piconet. Thus, the neighbor PNC is not a full member of the parent's piconet.

A DEV may decide to form a dependent piconet for a variety of reasons. If all of the channels are occupied with 802.15.3 piconets, other wireless networks, or with strong interferers, the DEV may want to share channel time with the existing piconet. The dependent piconet capability allows the DEVs to share the time on the channel while maintaining the synchronization required for the TDMA protocols in each piconet. Splitting the channel time between the two piconets allows them to achieve a higher level of coexistence. This coexistence capability is also available to other, non-802.15.3 wireless networks through the use of neighbor piconets.

Another reason for forming a dependent piconet is to extend the range of communications for a DEV. The parent PNC will only have a limited range, and this will limit the number of remote DEVs with which a given DEV can communicate. A DEV in the piconet may recognize that there are other DEVs with which it needs to communicate that are within range of the DEV but are not in range of the PNC. This DEV could then form a dependent piconet to communicate with those DEVs while still maintaining its connections in the parent piconet.

Security considerations can also be a reason for choosing to form a dependent piconet. In a piconet, all broadcast traffic is protected with the group key, and all of the DEVs in the piconet are able to hear this traffic. If a DEV wants to be able to restrict the distribution of this broadcast traffic, it could form a dependent piconet and only admit the DEVs to which it intends to send the broadcast data. Because the DEV would be the PNC of the dependent piconet, it can set the security policies to a level that is consistent with its applications. The DEV may also wish to form a dependent piconet if the parent PNC is not supporting security or if it using a security level that is not strong enough for the DEV's applications.

Once a DEV has decided that it wants to form a dependent piconet, it then needs to determine if it will form a child or a neighbor piconet. In general, the DEV will select to form a child piconet, but there are many reasons why a DEV would choose to create a neighbor piconet instead of a child piconet. One reason could be that the neighbor PNC is unable to enter into a secure membership with the parent PNC. An example of this is an apartment building where there are multiple piconets, each in a different apartment. In order to share the time on the air, the piconet of one apartment might want to get channel time from a nearby piconet, but it would be unable to authenticate to the parent PNC because the piconet did not belong to the same user as that of the parent PNC. Another reason for creating a neighbor piconet instead of a child piconet is that the neighbor PNC may be from another type of network, e.g., 802.11b, that only supports the minimal functionality required for a neighbor PNC. This DEV would be incapable of operating as a fully functional 802.15.3 DEV but might be able to support the commands required to join an 802.15.3 piconet as a neighbor and request channel time.

The most obvious difference between a child and a neighbor piconet is the DEVID that is used in the parent piconet. The PNC of a child piconet has a regular DEVID assigned to it in the parent piconet while the PNC of a neighbor piconet is assigned one of the reserved NbrIDs (see "Assigning DEVIDs" on page 80). The PNC of a dependent piconet is assigned a minimum of three different DEVIDs. The first DEVID is the one that is assigned to it in the parent piconet, either a regular DEVID or the NbrID. The second ID is the PNCID that it uses for command-related traffic in the dependent piconet of which it is the PNC. The dependent PNC also assigns itself a DEVID to use for data traffic in the dependent piconet, just as the PNC of the parent piconet has a DEVID assigned for data traffic.

Another key difference between a neighbor piconet and a child piconet is the frames that are allowed to be sent by the DEV in the parent piconet. Because the child PNC is a full member of the parent piconet, it can send data frames and any valid commands to any of the DEVs in the piconet. The neighbor PNC, on the other hand, is allowed to send only the following commands in the parent piconet:

- Association Request command
- Disassociation Request command
- Channel Time Request command
- Vendor Specific command
- Security Message command
- Any Probe Request, Probe Response or Announce commands
- Any required Imm-ACK frames

Because the neighbor PNC is a full member of its dependent piconet, it is allowed to send and receive data and commands during the operations of its piconet.

IEEE Std 802.15.3 does not require that all PNC-capable DEVs implement dependent piconet functionality. This capability is more complex than requirements for a regular PNC, and so the task group did not want to burden every PNC-capable DEV with this added cost. The PNC of a piconet is not required to allow neighbor piconets to be formed within its own piconet. The PNC is required to respond to a private channel time request from a DEV that could be used to start a child piconet, but it is not required to grant the channel time. The algorithm for granting channel time is outside of the scope of the standard, and the PNC may decline the request because it either cannot support pseudo-static allocations or the current traffic in the piconet is too dynamic to support the allocation of a pseudo-static CTA. This would have the effect of not allowing any child piconets without explicitly refusing them because the request is for a child piconet.

The initial proposals for creating dependent piconets used special commands for neighbor association and channel time requests. However, after working with dependent piconets in early stages of the draft, it was clear that some

---

simplifications could be made. First, the dependent piconet really only required channel time from the parent PNC. Other than that, the PNC did not need to be involved at all in the operations of the dependent piconet. The second attribute is based on the idea that in an 802.15.3 piconet once a DEV has been allocated channel time, it can use that time for essentially anything that it wants to do. The only restriction is that the DEV needs to ensure that its transmissions or the transmissions that it prompts, e.g., ACKs or allocations in a dependent piconet, do not extend beyond the temporal boundaries of the CTA. Thus, the dependent PNC can use the regular channel time request procedures to ask for time to be set aside for the operations of the dependent piconet. Finally, the idea of a pseudo-static CTA was used to assist the dependent PNC. Because this type of allocation is changed infrequently, it works well as the allocation for a dependent network.

STARTING A DEPENDENT PICONET

The first task in starting a dependent piconet is to associate with the PNC of the parent piconet. The DEV joins the piconet using the Association Request command (see "Association" on page 74) and follows the normal association process. If the DEV is going to become a child PNC, the association process is indistinguishable from the association of any other DEV. If the DEV is intending to become a neighbor PNC, there a few small differences in the association process.

First, the DEV sets the Neighbor PNC field to one in the PNC Capabilities field that it sends to the PNC in the Association Request command. When the PNC sees that this bit is set, it knows that the DEV is not requesting to become a full member of the piconet, but rather that it is asking to become a neighbor PNC. Because the PNC is not required to support neighbor piconets, the PNC can refuse the association request by sending an Association Response command with the Dest-ID set to the UnassocID, the requesting DEV's DEV address, and with the Reason Code set to "Neighbor piconet not allowed." If the DEV receives this type of a rejection to its association request, it should not try to associate with this piconet as a neighbor PNC as long as the parent PNC remains unchanged. The neighbor PNC DEV will know that the PNC has changed by examining the PNC Address field in the Piconet Synchronization Parameters field in the beacon.

The PNC may refuse the association for other reasons; for example, it may be in the middle of the PNC hand-over process or it may be changing channels. Depending on the Reason Code in the Association Response command, the DEV may wish to retry the request at a later time, for example, after the handover or channel change has completed. In general, the DEV can use the beacon to determine if the conditions have changed

 Notice The NbrID is not used for any data traffic in the parent piconet or for any data or command traffic in the neighbor piconet. During the time allotted for the neighbor piconet, the neighbor PNC uses either the PNCID or its DEVID in the neighbor piconet for communications. Thus the neighbor PNC has 3 DEVIDs: the NbrID in the parent piconet, the PNCID in the neighbor piconet and a regular DEVID in the neighbor piconet.

in such a way as to allow the DEV to associate as a neighbor PNC. For example, if the Reason Code is "Already serving maximum number of DEVs," then when the DEV sees a beacon with a DEV Association IE that indicates that a DEV has just disassociated, it may want to retry the association process with the PNC. If the PNC grants the association request for a neighbor PNC, it will assign the DEV one of the available NbrIDs to use for communications with the parent PNC.

When the association process is complete, either for a child or neighbor PNC, the next step is to request channel time for the dependent piconet. The DEV will use a regular Channel Time Request command with the following requirements:

- The Target ID will be the DEV's DEVID (either a regular DEVID or the NbrID).

- The Stream Index will be set to the Unassigned Stream Index (see "Stream connections" on page 92) This implies that the DEV is only able to request only an isochronous allocation as an asynchronous allocation would not make sense for this application.

- The PM CTRq Type field shall be set to ACTIVE. (A power-save mode allocation is not possible because the PNC is always required to be ACTIVE.)

- The CTA Rate Type field is set to indicate a super-rate allocation.

- The CTA Rate Factor is set to 1 to indicate one allocation per superframe.

- The CTA Type field is set to request a pseudo-static allocation.

All other fields in the Channel Time Request command can be set to values that are appropriate for that request. In particular, the dependent PNC will need to specify the minimum time that is required for it to support the operations of the dependent piconet.

If the SrcID for the CTRq command is one of the NbrIDs, the PNC will recognize it as a request for time for the neighbor piconet. If the request by a neighbor PNC is malformed in some manner, e.g., the Target ID is not its NbrID or the request is for a sub-rate allocation, then the PNC will reject the request. However, if the request comes from a regular DEV, the PNC will attempt to grant the request, even if it does not adhere to the above rules. The PNC will attempt this allocation because the child PNC is a member of the piconet, and so it is allowed to request channel time to other DEVs in the piconet or for purposes other than a dependent piconet. The PNC will consider each request based on the channel time available in the superframes, including the differential loading due to sub-rate and power-save mode allocations.

The PNC will send a Channel Time Response command to the DEV to indicate the result of the request for channel time. If the request is unsuccessful, the DEV can retry the request with different values for the timing parameters if it thinks this will improve the probability that the request will be granted. If the request is successful, then the PNC will begin to place the CTA blocks in the beacon to allocate the time for the DEV to operate as the dependent piconet. At this point, the DEV is now the PNC of the dependent piconet and it is allowed to begin sending a beacon, associating new members and allocating channel time. The dependent PNC must select a PNID that is distinct from the parent piconet's PNID. If the dependent PNC did not do this, then DEVs in the parent piconet that were in receive mode during the dependent piconet's CTA might ACK frames or respond to commands that were intended for a DEV in the dependent piconet that had the same DEVID as the DEV in the parent piconet. The process of a DEV becoming a child PNC is illustrated in Figure 5–1.

If the dependent piconet is a child piconet or an 802.15.3 neighbor piconet, then the dependent PNC will need to set aside time in the beacon that corresponds to the time in the parent piconet that is not allocated to the dependent piconet. The child or 802.15.3 neighbor PNC does this by putting a CTA block in the beacon with the SrcID and DestID set to the PNCID and the start time and duration corresponding to the time that is reserved for the parent piconet. Because 802.15.3 has strict rules regarding when a DEV is allowed to transmit, the child PNC is

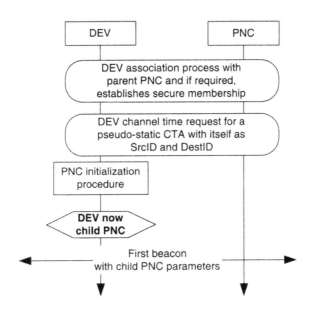

Figure 5–1:   Process for starting a child piconet

able to ensure that all of the DEVs in its piconet do not violate the time allocated for the dependent piconet operation. The alignment of the child and parent superframes is illustrated in Figure 5–2.

Figure 5–2 shows that in a child piconet, the start time of the reserved channel time is equal to the end of the allocation for the dependent piconet. Likewise, the end of the reserved channel time corresponds to the start

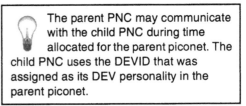

The parent PNC may communicate with the child PNC during time allocated for the parent piconet. The child PNC uses the DEVID that was assigned as its DEV personality in the parent piconet.

time of the CTA for the dependent piconet. Because the child is a member of the parent piconet, there are times during the parent superframe when the child PNC can communicate with the parent PNC or with other DEVs in the parent piconet. This is indicated in Figure 5–2 as blocks of time marked with "C-P." When either of the PNCs, child or parent, is sending a beacon, communication is not possible (other than the information passed in the beacon). The parent PNC is also not allowed to communicate with the dependent PNC during the blocks of time marked "C-C." Although the child PNC is a member of the parent piconet, the parent PNC is not necessarily a member of the child piconet, and so it cannot communicate with the other DEVs in the child's piconet. The parent PNC may join the child piconet if it wants to communicate with DEVs that are associated in that piconet but are not associated in the parent piconet. If the parent PNC joins the child piconet, it will be as a regular DEV and not as the PNC.

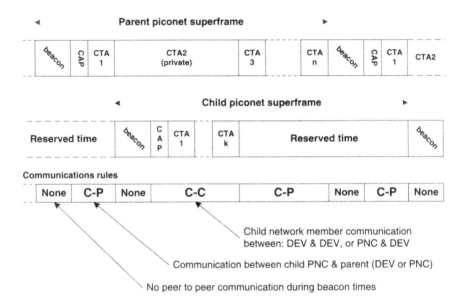

Figure 5–2:   Relationship of the parent superframe to the child superframe

As Figure 5–2 shows, the child PNC can act as a bridge between the two piconets, the parent and the child piconet. The child PNC is allowed to implement either a bridging protocol, such as 802.1d, or a routing protocol to enable data to be passed between the two piconets. However, these protocols are functions that exist at the higher protocol layers and so are out of the scope of IEEE Std 802.15.3 MAC and PHY. An example of this bridging is illustrated in Figure 5–3.

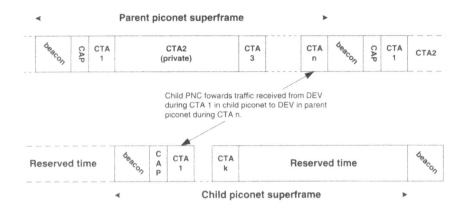

Figure 5–3: Child PNC bridging data between networks

The process for a DEV becoming a neighbor PNC is illustrated in Figure 5–4. As the figure indicates, the neighbor DEV may not be operating an 802.15.3-compliant network. However, if the DEV wants to join as a neighbor PNC, it is required to support the minimal functionality described in IEEE Std 802.15.3. In addition, the neighbor PNC must make sure that all of the members of its wireless network do not transmit outside of the time allocated for the dependent piconet. If the neighbor PNC is unable to guarantee this, it is not allowed to join the 802.15.3 piconet as a neighbor PNC.

The relationship between the superframes of the parent piconet and the neighbor piconet are illustrated in Figure 5–5. The parent and neighbor PNCs can carry on limited communications during the time periods marked "N-P," although this communication is restricted to the commands necessary to maintain the neighbor piconet. The neighbor PNC is not allowed to communicate with DEVs in the parent piconet because it is not a member of that piconet. Likewise, the parent PNC is not allowed to engage in any frame exchanges during the time set aside for the neighbor piconet because it is not a member of that piconet. As with the case of the child piconet, the parent PNC is allowed to join the neighbor piconet so that it can communicate with the DEVs in that network (assuming it supports the protocol that is being used in the neighbor piconet).

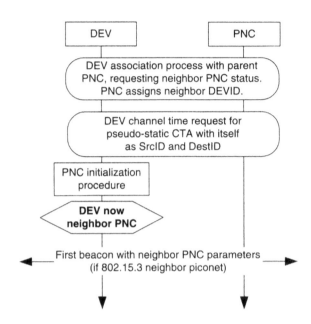

Figure 5–4: Process for starting a neighbor piconet

PARENT PNC CEASING OPERATIONS WITH DEPENDENT PICONETS

When the parent PNC stops the parent piconet using the shut down procedure, the dependent piconets will need to adjust for the loss of their synchronization. If there is only one dependent piconet that is part of the parent piconet, that piconet will be able to continue operations. However, if there is more than one dependent piconet that is a part of the parent's piconet, then they both cannot continue to operate. Once the parent PNC stopped sending beacons, the dependent PNCs might assume that they can use the entire time in the channel, expanding to fill the time that was allocated for traffic in the parent piconet. However, if all of the dependent piconets do this, they would collide, causing lost frames and poor QoS.

One way to resolve this problem would be to require that all dependent piconets cease operations if the parent piconet also ceases operations. However, it is possible for one dependent piconet to remain as long as the others know to cease operations. The task group solved this problem by adding a field to the Piconet Shutdown IE that contains the DEVID of the dependent piconet that is allowed to

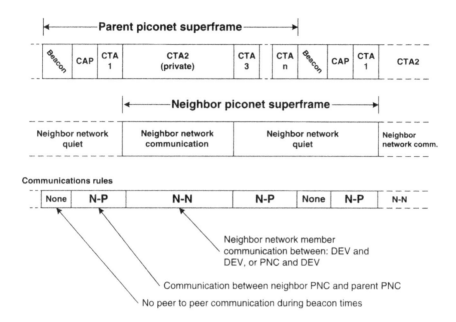

Figure 5–5: Relationship of the parent superframe to the neighbor superframe

continue operations. Rather than forcing the PNC to determine which dependent piconet is most deserving to continue operations, the standard requires the PNC to select the dependent piconet with the lowest DEVID to continue operations. It is unlikely that the PNC could determine the "best" piconet from its limited knowledge and in the limited amount of time allowed for the piconet shutdown procedure. Thus, an arbitrary choice is as good as an attempt to select the "best" one, and it requires very little overhead in an implementation. Using the lowest DEVID also gives preference to child piconets over neighbor piconets because the values of the NbrIDs are all greater than the highest allowed DEVID value.

All of the other dependent PNCs, if they do not see their DEVID in the beacon, will be required to either shut down their piconets, change PHY channels, or join another piconet as a dependent piconet before the parent PNC sends its last beacon. Once the parent PNC has sent its last beacon, all DEVs that are not members of the dependent piconet that was selected to continue operations will have to cease all transmissions. The dependent PNC that was selected to continue operations will remove the Parent Piconet IE from its beacon to signify that it is no longer a dependent piconet and it can likewise remove the allocation in its superframe for the parent piconet's operations. DEVs that were a part of any of the piconets that shut down, either from the parent piconet or from another dependent piconet, can then join the remaining piconet either as DEVs or to start a dependent piconet.

PARENT PNC STOPPING A DEPENDENT PICONET

The parent PNC is allowed to force a dependent piconet to cease operations. The method that the parent PNC uses to do this depends on the type of dependent piconet that it wants to shut down. If the parent PNC is going to shut down a child piconet, it uses the stream termination procedure (see "Stream connections" on page 92) to remove the channel time that is allocated for the child piconet. The parent PNC sends the Channel Time Response command to the child PNC with a Reason Code that explains the reason that the parent PNC is removing the CTA. The child PNC remains a member of the parent piconet, and its other stream and asynchronous connections remain unchanged by the procedure.

If the dependent piconet is a neighbor piconet, the parent PNC uses the disassociation process (see "Disassociation" on page 78) to end the neighbor PNC's association with the parent piconet. The Disassociation Request command is used instead of the stream termination procedure because the neighbor PNC's only reason for associating with the parent PNC is to operate a neighbor piconet. If the parent PNC is not going to allow the neighbor piconet to operate, there is no reason for the neighbor PNC to remain associated in the parent piconet.

Once the dependent PNC has been informed, either via the Channel Time Response command or the Disassociation Request command, that it will no longer be able to operate a dependent piconet as a part of the parent piconet, the dependent PNC can then either shut down its piconet, change channels, or join another piconet as a dependent piconet. The parent PNC will listen for the dependent PNC's shut down or channel change procedure to determine when to remove the channel time for the dependent piconet from the beacon. However, the parent PNC is not required to wait for an indefinite period of time and it may choose a timeout for the dependent piconet that is shorter than the time indicated in the dependent PNC's shut down or channel change process. If the dependent piconet is an neighbor piconet that is not 802.15.3-compliant, the parent PNC will select a time that is equivalent to what it would use to shut down its own piconet to allow the dependent piconet to either shut down or change to another channel.

Regardless of the length of time that the PNC allows for the dependent PNC to cease operations in the current PHY channel, when the CTA for the dependent piconet is removed from the beacon, all of the DEVs in the dependent piconet, including the dependent PNC, shall cease transmitting in that CTA. Because the CTA for a dependent piconet is a pseudo-static allocation, the parent PNC might want to allow mMaxLostBeacons number of superframes before it reallocates that time to another DEV. It can allocate it sooner, but the first few superframes may experience high frame loss if the DEVs in the dependent piconet have not yet realized that the allocation has been finally removed.

HANDING OVER PNC RESPONSIBILITIES IN A DEPENDENT PICONET

One of the key capabilities that IEEE Std 802.15.3 provides is the ability of the PNC to hand over control of the piconet to another DEV in the piconet. This capability is critical to the 802.15.3 protocol because the "best" PNC is not initially selected to be the PNC of the new piconet. Although early drafts of the standard included a PNC election process, this procedure was dropped in favor of simply letting the first PNC-capable DEV on the air become the PNC. This DEV will then hand over control to the "better" DEVs as they join the piconet. Eventually, the same result as a PNC election process will occur, in that the "best" DEV becomes the PNC of the piconet. PNC hand-over also allows the piconet to continue uninterrupted when the PNC needs to leave the piconet for any reason.

Thus, the PNC hand-over process is a key part of the operation of IEEE Std 802.15.3. However, up until the first sponsor ballot, the PNC hand-over process for dependent PNCs was not even supported in the standard. This capability was added after the first sponsor ballot and now is a part of the standard for DEVs that implement the dependent PNC capability.

Although the parent PNC can hand over PNC responsibilities to a DEV that is currently the PNC of a child piconet, this does not imply that the dependent PNC will then merge the two piconets. The standard does not provide either the commands or the procedures that are necessary to support merging two piconets.

This may be a future enhancement to the standard, but it is not currently supported. The PNC is never allowed to hand over control to a neighbor PNC because this DEV is not a full member of the parent piconet. If the parent PNC does hand over control of the piconet to a DEV that is currently a child PNC, that DEV would have to operate two distinct piconets simultaneously. If the child PNC is unable to support this mode of operation, it will refuse the hand-over request by sending the PNC Handover Response command to the PNC with the Reason Code "Handover refused, unable to act as PNC for more than one piconet."

Handing over control of dependent piconet required the addition of two new capabilities to the standard. The first capability is to allow a DEV, under certain circumstances, to refuse a hand-over request. In almost every other case[28], the DEV that is selected by the PNC to become the new PNC is required to accept the nomination and will become the new PNC of the piconet. However, in the case of a dependent piconet, the new dependent PNC will need to become a member of the parent piconet before it can take over as the PNC of the dependent piconet. This may not be possible for any one of a number of reasons. For example, the DEV may be unable to enter into a secure relationship with the parent PNC, and so it cannot become a member of the parent piconet. This issue does not apply to neighbor piconets where the neighbor PNC is only associated and is not required to have a secure relationship with the parent PNC. The selected DEV may not be able to associate with the parent PNC because the parent PNC is unable to handle any new DEVs in its piconet. The selected DEV may not even be in range of the parent PNC; in which case, it will be impossible for it to operate as the dependent PNC. If the selected DEV is unable to join the parent piconet, the DEV will refuse the hand-over process by sending the PNC Handover Response command to the PNC with the Reason Code set to "Handover refused, unable to join parent piconet."

The other change that was necessary in the standard was to allow a DEV to hand over its CTA to another DEV. Assuming that the selected DEV is able to join the parent piconet, it will be assigned a DEVID that is different from the current dependent PNC. The CTA for the dependent piconet, however, will be associated with the old DEVID and the new PNC will be unable to modify or respond to any

---

[28] The other exception is that a DEV that is acting as a child PNC may refuse the handover due to its inability to run more than one piconet at the same time.

changes in the allocation. Even worse, if the current dependent PNC is handing over control because it is going to shut off, at some point, it will disassociate from the parent piconet and the parent PNC will then remove the CTA from the beacon and the dependent piconet will cease to exist. Thus, the parent PNC needs to be informed that a new DEV will be taking over as the "owner" of the time allocated to the dependent piconet.

The hand over of channel time is restricted to private pseudo-static allocations only, and only the current source of the allocation can request a change in the "ownership" of the CTA. To hand over the CTA to another DEV, the source DEV sends a Channel Time Request command to the PNC that contains the following information:

- The Num Targets field is set to one.
- The Target ID List field contains the DEVID of the target DEV that is to receive control of the CTA.
- The Stream Request ID field is set to zero.
- The Stream Index field set to the stream index of a CTA that has already been allocated to the dependent PNC as a private, pseudo-static CTA.
- All other fields set to the same values as in the last successful CTRq for this Stream Index.

The PNC will notice a change in the Target ID List and will know that this is a request to hand over control of the CTA to the DEV indicated in the Target ID list. The PNC will grant this request as long as the DEV that is to receive control of the CTA is a mem-

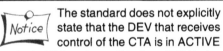

The standard does not explicitly state that the DEV that receives control of the CTA is in ACTIVE mode, but it is not possible for a DEV to be in a PS mode and have a super-rate allocation. Thus the DEV needs to be in ACTIVE mode when the control is passed.

ber of the piconet and its power management mode is ACTIVE. The PNC will respond with the Channel Time Response command indicating the success or failure of the request. If the request failed, the PNC will set the Reason Code field to indicate the reason why the request was refused. The frame exchange sequence for handing over control of a CTA is illustrated in Figure 5–6.

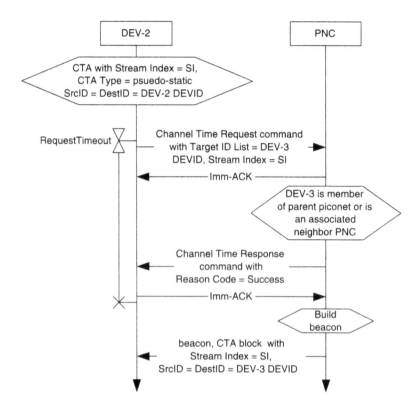

Figure 5–6: Dependent PNC handing over control of a CTA

With these two capabilities, the standard defines a procedure that will allow a dependent PNC to hand over control to another member of its piconet. Perhaps more important, it is possible to describe the potential failures in the dependent PNC handover process and how the protocol can handle them.

The dependent PNC hand-over process begins in the same manner as a regular PNC hand-over, with the current PNC sending the PNC Handover Request command to the DEV that it has selected to become the new PNC, as shown in Figure 5–7. In this and the two subsequent figures, the identities PNC, DEV-2, and DEV-3 are all relative to the dependent piconet and not the parent piconet. If DEV-2 is not a member of the parent piconet, then the DEV will begin the association process to join the parent piconet and, if required, establish a secure relationship with the PNC. DEV-2 can join the parent piconet as either a neighbor or as a member of the piconet. While DEV-2 is attempting to join the parent piconet, the current dependent PNC will be sending DEV-2 the information about all of the DEVs with the PNC Information command, all of the current channel time requests with the PNC Handover Information command, and the power save information, if any, using the PS Set Information Response command. DEV-2 may also request the transfer of any security information at this point using the Security Information Request command. The dependent PNC transfers all of this data, anticipating that DEV-2 will be successful in joining the parent piconet. The transfer of this information does not interfere with either DEV-2's association and secure membership process or with the transfer of the dependent piconet data because the former occurs only during time reserved for the parent piconet, whereas the latter only occurs during time reserved for the dependent piconet.

Once the transfer of the information is complete and DEV-2 has joined the parent piconet, it will send a PNC Handover Response command to the dependent PNC with a Reason Code set to the DEVID that was assigned to it by the parent PNC. This lets the current dependent PNC know that the new dependent PNC is ready to take over control of the piconet. The Reason Code in the PNC Handover Response command is interpreted as follows:

- *0x00:* Success. (This code is used only for a non-dependent PNC handover.)
- *0x01–0xEC:* Success, the DEV that is to become the new dependent PNC is a member of the parent piconet with a DEVID equal to the value of the Reason Code. (These values cover all possible DEVIDs for members of a piconet.)
- *0xED–0xF6:* Reserved. (These values are reserved DEVIDs and so they are reserved for this command as well.)
- *0xF7–0xFC:* Success, the DEV that is to become the new dependent PNC has associated in the parent piconet as a neighbor PNC with a DEVID equal to the value of the Reason Code. (These values are all NbrIDs.)
- *0xFE:* Handover refused, unable to join the parent piconet. (This value is the same as the unassociated ID, which is appropriate for this error code.)
- *0xFF:* Handover refused, unable to act as a PNC for more than one piconet. (This value is the same as the BcstID and so it is never assigned to a DEV as its DEVID).

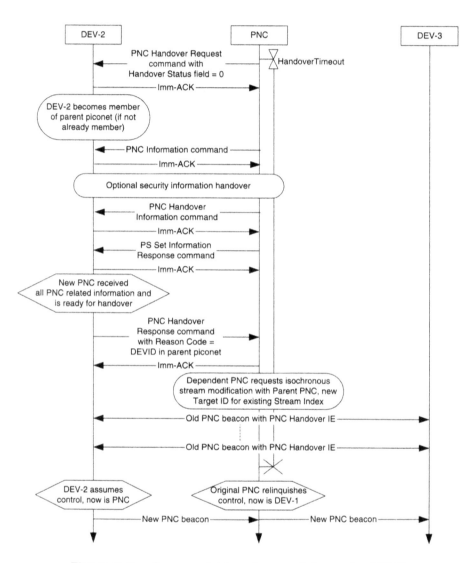

Figure 5–7: Successful hand over of dependent PNC

At this point, the dependent PNC uses the Channel Time Request command to hand over the control of the dependent piconet CTA to the new dependent PNC. Once the parent PNC changes the SrcID and DestID of the dependent piconet CTA, the current dependent PNC must complete the hand-over process because it will not be able to regain control of the CTA.

After the control of the dependent piconet CTA has been transferred to the new dependent PNC, the current dependent PNC will begin placing the PNC Handover IE in its beacon with the Handover Beacon Number field set to the beacon number of the first beacon that will be sent by the new PNC. The last superframe controlled by the current dependent PNC will be the one in which the beacon number is one less than the value of the Handover Beacon Number field. The following superframe will begin when the new dependent PNC sends its first beacon.

There are multiple points in the hand-over process where it can fail. The current dependent PNC can cancel the hand-over process up until the time that it requests that the parent PNC hand over control of the dependent piconet CTA to the new dependent PNC. The dependent PNC cancels the process by sending a PNC Handover Request command to DEV-2 with the Handover Status field set to 1 to indicate that the process has been cancelled.

The hand-over process can also fail if DEV-2 fails to join the parent piconet. If DEV-2 attempts to join as a neighbor PNC and the parent PNC does not support neighbor PNCs or does not wish to allow any more neighbor PNCs, then the association request by the new dependent PNC will be rejected. In that case, the DEV can also try to join as a regular DEV; in which case, the dependent piconet would become a child piconet after the hand-over process. Although it may seem strange that the hand-over process can change a piconet from being a neighbor piconet to being a child piconet, it is important to remember that the difference between a child and a neighbor piconet lies only in the characteristics of the dependent PNC. If the dependent piconet is currently a child piconet and DEV-2 attempts to join as a neighbor, the parent PNC will be able to decide if it wants to support the piconet as a neighbor piconet. If the parent PNC does not want to allow a neighbor piconet, then it will reject the association request of the DEV as a neighbor PNC. Likewise, if the dependent piconet is currently a neighbor piconet and DEV-2 either is a member or becomes a member of the parent piconet, then the parent PNC should allow the piconet to become a child piconet.

The PNC has already allowed DEV-2 to become a member of the parent piconet, so it should not have any problem with DEV-2 taking over control of the dependent piconet's CTA. Once the CTA has been allocated, there is minimal additional effort required by the PNC to change the SrcID and DestID for the CTA.

If DEV-2 fails to join the parent piconet as either a regular DEV or as a neighbor PNC, it will send the PNC Handover Response command to the dependent PNC with the Reason Code set to 0xFE, which is an error code that means "Handover refused, unable to join parent piconet." Note that the error code is equal to the UnassocID, which is DEVID that the DEV-2 will receive if it is unable to associate with the parent PNC. The frame exchange leading up to the failure of DEV-2 to join the parent piconet is illustrated in Figure 5–8. As shown in the figure, DEV-2 is allowed to refuse the handover at any time while the dependent PNC is sending over the information about the piconet. DEV-2 does not have to wait until all of the information is transferred before it sends the PNC Handover Response command if it is going to reject the handover.

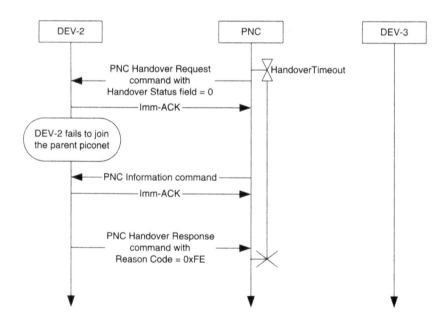

Figure 5–8: Hand-over of dependent PNC fails because new PNC fails to join the parent piconet

If DEV-2 is successful in joining the parent piconet, then the next problem that is unique to dependent PNC hand-over is handing over control of the CTA to the new dependent PNC. Because DEV-2 has passed its new DEVID to the current dependent PNC in the PNC Handover Response command, the current dependent PNC should not make a mistake in selecting the correct DEVID to send in the Channel Time Request command. However, the PNC may reject the handover of the CTA control if any of the following are true:

- The DEVID specified in the target ID list does not correspond to an associated DEV.

- The DEVID specified in the target ID list does not correspond either to a secure member of the piconet or a NbrID and secure membership is required in the parent piconet.

- The DEVID specified in the target ID list is in power-save mode.

- The PNC is terminating the stream.

The PNC should have time available to allocate the CTA because it is already allocating the same amount of time for the dependent piconet. If the parent PNC rejects the request to hand over control of the CTA to the new DEV-2 PNC, the dependent PNC will send a PNC Handover Request command to DEV-2 with the Handover Status field set to 1 to indicate that the hand-over process is being cancelled, as illustrated in Figure 5–9.

If the dependent PNC cancels the hand-over process, DEV-2 may choose to disassociate from the parent piconet because there is no longer a need for it to be a part of the parent piconet. If DEV-2 joined as a neighbor PNC, it should disassociate from the parent piconet when the hand-over process is cancelled to free up that resource for other DEVs that may want to form a neighbor piconet.

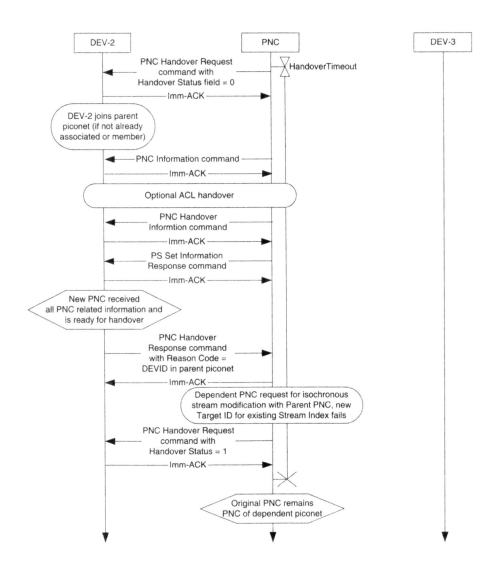

Figure 5–9: **Hand-over of dependent PNC fails because of failure to hand over control of the dependent piconet CTA**

Chapter 6 Security

Wireless networks are by their nature a broadcast medium. They provide no intrinsic PHY level security, at least for standardized solutions. Occasionally, people will mistake the use of a technology by the military, e.g., spread-spectrum or ultra-wideband, to indicate that the PHY has some sort of built-in security. However, the military does not publish interoperability guides to its secure systems, whereas that is exactly the purpose of a standard. Standardized PHYs allow anyone reasonably skilled in the art to successfully extract the data from the wireless medium. If fact, many wireless networking cards have a "promiscuous mode" wherein the radio demodulates all of the frames on the air and passes them up for further analysis.

Manufacturers and people in the press have erroneously claimed from time to time that standardized solutions for CDMA (IS-95) [B28], DSSS (802.11b) [B13], FHSS (802.11 and RangeLAN) [B12] or even ultra wideband (UWB) PHYs offer some sort of security inherent in the PHY layer signaling.[29] The reality is that none of these PHYs intrinsically offer even a modest level of security. Instead, the only strong security available to a wireless network is the tried and true method of using bit-level encryption techniques on the data. This is actually good news because the security provided by these techniques is well studied and understood by experts in the field. Very strong security can be achieved with a minimal impact in terms of logic gates and power in an IC implementation, even at data rates as high as 500 Mb/s.

INTRODUCTION AND HISTORY

The security in 802.15.3 is based on three capabilities:

a) *Authentication and authorization:* The ability to determine the identity of a DEV and decide if it should be allowed in the piconet.

---

[29] Quantum cryptography that can actually detect when the PHY layer signaling is modified is an exception this rule. It may be that PHYs based on these techniques will offer security that is inherent in the PHY signaling.

b) *Data integrity:* The ability to determine the true source of the data and to verify that the data was not changed once it had been sent.

c) *Data privacy:* The ability to prevent unauthorized DEVs from gaining access to data that is sent over the air.

The latter two capabilities are provided by the symmetric key encryption suite (see "Symmetric key security suite" on page 195), which provides MAC level encryption and/or integrity checking of the data.[30] The authentication and authorization portion of the draft caused the biggest problem with the development of the draft as well as some of the most contentious debates at meetings and on the e-mail reflector.

> The final draft of the standard was numbered D17 and so one might surmise that there were 17 released versions of the draft. In fact, there were 21 released versions of the draft. Following D09, the draft versions adopted hexadecimal notation. This provided some humor as the next draft was appropriately named D0A (i.e., Dead On Arrival). This was followed by D0B, D0C, D0D and then D10 and the rest followed numerically. (Versions D0E and D0F were skipped.)

The initial draft released for working group ballot, D09 (i.e., the ninth version of the draft), did not include any security suites to enable authentication (or even encryption) but rather included variable length frames that could be used by a security process, e.g., some form of public-key cryptography, to implement authentication. The actual process for performing the authentication was determined to be out of scope.[31]

However, in the first working group ballot, a sizeable percentage of the voters (about 12%) requested that the draft include a specific authentication process to enable interoperability among consumer electronic devices. To resolve the comments of these voters, the task group began a process of a call for proposals and a downselection process to select a single public-key cryptography security suite and symmetric key suite for inclusion in subsequent versions of the draft. Two public-key suites were proposed, one based on elliptic curve cryptography (ECC) and digital certificates proposed by individuals from Certicom™ and Motorola® and another one based on the short lattice vector problem proposed by

---

[30] The Integrity Code field in 802.15.3 is normally referred to as a message authentication code or MAC in cryptographic circles. However, the acronym MAC is already being used by the standard in a few places, so IEEE Std 802.15.3 refers to this field as the IC instead.

[31] This was in November 2001; this same conclusion was to be reached in January 2003 when the draft comes full circle on this issue with draft D16.

individuals from Ntru™. In addition, two symmetric key security suites were proposed, the advanced encryption standard (AES-128) [B22] and another based on triple data encryption standard (triple-DES).

The selection of the symmetric key suite was relatively straightforward and uncontentious, and AES-128 was selected to provide the symmetric key services. Likewise, the group felt strongly that digital certificates should be an option and not a requirement for the standard. The public-key security suites resulted in a much more difficult selection process. The task group was concerned because the encryption algorithm used by Ntru was patented as were certain protocols often used to implement ECC public-key cryptography. With the possibility of required royalty payments on the line for both sides, the task group selected ECC but failed to get the 75% approval[32] required to go forward. The group worked out a compromise; the technical editor would put in an ECC security suite that had the least amount of essential IP into the next draft as the mandatory security suite, and Ntru would be included as an optional security suite. A due date was provided for the submission of the ECC suite, and if a complete solution was not provided to the technical editor by this date, the Ntru security suite would become the mandatory security suite.

Two proposals for an ECC security suite were received by the deadline, and a working group letter ballot had to be held to determine the one that would be used in the draft, delaying the release of the draft and the start of another working group letter ballot for at least two months. Based on the result of the working group ballot, one of the two ECC suites was put into the draft as mandatory for the next recirculation ballot. This, predictably, resulted in more no votes that had to be resolved. Finally, a compromise was struck. Three public-key security suites would be provided, ECMQV Koblitz-283[B27], NTRUEncrypt 251 [B3], and RSA-OAEP 1024 [B4], all optional for a DEV to implement. The only restriction was that if a DEV implemented security, it had to support at least one of the three security suites.

Although it seemed that this would finally settle the question of security suites in the standard, it came to light during sponsor ballot that authentication and authorization were indeed outside of the scope of a MAC and a PHY. In fact, the

---

[32] The motion failed by one vote.

security suites made mention of the fact that authorization and even some parts of the authentication protocol, such as digital certificates, were outside of the scope of the standard. Rather than risk the standard being sent back from either the IEEE 802 Executive Committee or the IEEE-SASB Standards Review Committee (RevCom), the task group decided to remove the authentication portion of the security suites and retain only the frame formats with variable length fields that would support any type of authentication protocol.

This exercise was not entirely futile. The process of selecting a security suite provided a focus on the effects of security on the piconet. The process also helped to solidify the types of frames required for authentication as well as defining the methods used to update keys in a secure relationship. IEEE Std 802.15.3 also added peer-to-peer security to the group level security that was first proposed. Finally, IEEE Std 802.15.3 led the way by adopting counter plus chaining block cipher message authentication code (CTR + CBC-MAC or CCM) mode for AES-128 using a single key. This was later adopted by both 802.15.4 and 802.11i as the mandatory encryption algorithm for those standards. Overall, the process of selecting, reviewing, and evaluating the security suites helped to improve the overall quality of the security services in the 802.15.3 draft standard. It also provided an opportunity for the members of the 802.15.3 task group to become better educated on the implementation considerations for security.

> One of the consequences of purging the security suites from the the standard is that the security terminology was changed. For example, anywhere that the standard says "secure membership" it really means "authenticated and authorized." Likewise, the "key originator" is really the "security manager." The Security Message command is intended to be used for 3 and 4 pass authentication protocols and the Security Information command is intended to carry the ACL and the associated authentication information.

SECURITY MODES AND POLICIES

The 802.15.3 standard operates in one of two security modes, mode 0 and mode 1. When the piconet is operating in mode 0, no security is provided and the piconet is wide open. In this mode, individual DEVs may form secure relationships or the higher layers may form secure links, e.g., a virtual private network (VPN), to protect the privacy and integrity of the data. Piconet-wide information such as the beacon or commands sent to or by the PNC are not protected in any fashion in mode 0. In mode 1, all commands and the beacon have their integrity protected and all data are encrypted to provide privacy and have their integrity protected. All of the DEVs in the piconet have access to same group key, so each DEV in the piconet will have to evaluate if this level of security is acceptable for its requirements.

The security model for 802.15.3 has a DEV that acts as the key originator (KO) for other DEVs in that security relationship. The key originator keeps track of the keys and forces an update to them whenever required by a security event. When the piconet is operating in mode 1, the PNC acts as the key originator for that group security relationship.

Security services provided in mode 1

The following security services are provided in mode 1:

a) *Key transport:* The key originator in a relationship can update the symmetric keys that are used, and the DEVs in the relationship can also request new keys if they detect that the encryption keys have changed.

b) *Data encryption:* This service provides privacy for the data that are transmitted in the piconet. If the piconet group key is used, then every DEV in the piconet will be able to read and understand the data. The data can also be protected using a key that is only shared between two DEVs.

c) *Data integrity:* Frames that require data integrity use an Integrity Code field that is part of the frame format to protect the data from modification and to verify the source of the data.

d) *Beacon integrity protection:* The information in the beacon is not encrypted because it is used by DEVs that are searching for piconets to join. The information in the beacon is protected with an Integrity Code field to protect the information from modification and to verify that it was sent by the PNC.

e) *Freshness protection:* Every beacon sent by a PNC includes a time token that is used by the encryption algorithm as one of the inputs used to create the nonce. The time token is included in the beacon regardless of the security mode. This enables peer-to-peer security in a piconet that is operating in mode 0.

f) *Command integrity protection:* Like the beacon, commands sent in a piconet operating in mode 1 include an Integrity Code field to verify that the command has not been modified and to verify the identity of the source of the command. Most of the commands sent in the piconet are protected. Exceptions to this are for commands that are sent before a DEV can obtain secure membership, e.g., Association Request command, Association Response command, and Disassociation command.

The following security services are outside of the scope of IEEE Std 802.15.3 and so are not provided with mode 1:

a) *Access control list:* The access control list provides a method for the key originator to determine which DEVs are authorized to operate in the piconet.

b) *Mutual authentication:* This process verifies the identity of the parties in a security transaction. This process will provide the management keys required in the security relationship to be able to update the symmetric keys.

c) *Key establishment:* Generally, this will be a part of the authentication process. In the process of authenticating with another DEV, the DEVs will create a shared secret that can then be used to generate the management keys for the relationship. Because this is tied to the authentication process, it is outside of the scope of the standard.

d) *Authorization:* This is one of the most important security services and it is normally provided by the end user at the application level. In certain applications, e.g., a home theater system, it may be possible for the manufacturer to pre-program the authorizations required to enable secure operation of the DEVs.

The security service listed in item d), authorization, is one that has never really been within the scope of the standard. Although it is possible to provide automatic methods to authenticate a DEV using digital certificates, this still does not provide any information regarding authorization. For example, a person purchases an extra pair of wireless speakers for their home audio system. When the user turns on the speakers, they provide a digital certificate to the audio system that proves that they were produced by "Manufacturer A" and provides some sort of identification, i.e., the MAC address. However, it is not possible for the audio system to know that these speakers are indeed the speakers that the person just picked up at their local electronics store and are not, for example, the speakers in the next-door neighbor's house that happen to be made by the same manufacturer. At some point, a person will need to intervene in some fashion to declare that the new DEVs are either allowed in the piconet or that they will be refused membership.

There are a variety of methods that can be used to authorize a DEV for operation in a given piconet. In a cordless phone, the handset often uses the physical contacts of the battery charger to send information that authorizes the handset to work with that base station. This is a sufficient level of security because if the malicious user could gain physical control of both the handset and the base station, the physical security of the home would have already been severely compromised. Another method for authorization is for the audio system to display a unique number sent by the new DEV that is also provided in the documentation for the product. The user just needs to verify that the number matches what is printed on the product in order to authorize the DEV.

Security policies

An important part of the 802.15.3 security architecture is the security policies that are required to protect the information in the piconet. These policies dictate specific security-related actions in response to a security event in the piconet. One of the key security policies is to make sure that the DEVs that are supposed to be members of the security relationship are still participating in the piconet. This is similar to the ATP being used to determine the DEVs that are still participating in the piconet. If the key originator has not received a frame of some sort from the DEV within that time period, it will deauthenticate the DEV. If the key originator is the PNC, the time period is the ATP and the result of a time out will be a disassociation, which also includes a logical deauthentication.

If the membership of the group changes, i.e., a DEV joins or leaves the group, the key originator is required to change the symmetric keys. The KO signals this change by changing the security ID (SECID) used in its frames and then uses the Distribute Key Request command to send the new keys to the DEVs in this security relationship. The SECID is used in the frame to indicate the key that is to be used to either check the data integrity or to decrypt the frame. If the KO is the

PNC, then the events that change the group membership are the association and disassociation procedures. Every time that the PNC either associates a new DEV or disassociates an older one, it must change the keys and the SECID.

PNC handover also affects the makeup of the security relationship because the PNC is also the key originator for the piconet. After the PNC passes over control to the new PNC, all of the DEVs in the piconet will be required to establish a secure relationship with the new PNC. The DEVs need to do this so that they can get a copy of the management key that will be used for subsequent key exchanges. If the DEVs that are already members of the piconet have entries in the ACL, then the new PNC can use the ACL to determine if a DEV should be added to the piconet without necessarily needing input from the user.

As a part of the 802.15.3 security policies, there is a list of all of the possible frames and the keys required to send these frames. The two-and-a-half page table is not replicated here, but the rules in the table need to be followed in order to maintain the security of either the piconet or the security relationship. As mentioned earlier, some of the command frames are always sent with security turned off. In addition, a DEV may send a data in a non-secure data frame while the piconet is operating in Mode 1. One of the potential applications for this is streaming video. The content providers will likely provide end-to-end encryption of the data stream that starts and ends at the application layers. Encrypting this data, which are already encrypted, is a waste of clock cycles and octets in the frame. Instead, the DEV can send these data frames with security turned off to enhance the efficiency of the network.

When a DEV receives a secure beacon, it will first check to see that the time token in the beacon is strictly greater than the previous valid time token and is less than that value plus mMaxTimeTokenChange, a parameter with value of 65,535. The DEV will accept time tokens in that range because it may have missed some of the beacon frames due to interference or because the DEV was sleeping. If the SECID matches the DEV's current SECID for the piconet group key and the beacon did not generate an integrity check failure, then the DEV will update its value for the last valid time token. Because the time token is used to preserve the freshness of commands in the piconet, it is very important that the DEV not get too far out of sync with the PNC in its expectation for the time token number.

A DEV performs similar checks for the command frames that it receives, except that it uses the time token that is correct for the current superframe, even if the DEV did not correctly receive the beacon. The time token is sent only in the beacon and is not included in the secure frame fields.

SYMMETRIC KEY SECURITY SUITE

The data privacy and data integrity capabilities in this standard are provided by a symmetric key encryption suite. The encryption algorithm used is AES with 128-bit keys and 128-bit blocks. The AES algorithm is the result of a worldwide competition and downselection for a new encryption standard for the U.S. that could be used across a variety of applications. Accordingly, the algorithm was analyzed not just for its security properties, but also for the efficiency of its implementation. The AES algorithm has been subjected to intense scrutiny by the cryptographic community and is considered to be a very reliable algorithm.

However, as important as the algorithm is, equally important is the mode in which the algorithm is used. AES is often proposed to be used in open code book (OCB) mode. AES OCB has been proven to be secure, but it is also a patented mode of operation. The 802.15.3 Task Group was concerned with the potential cost of a patented solution, so they pursued a mode of operation that is thought to be IP free.[33] This led to the selection of the counter plus chaining block cipher message authentication code (CTR + CBC-MAC or CCM) mode.

Overview of AES CCM

The AES CCM mode is described in detail in the standard, but a few key features will be discussed here. The CCM mode uses the same encryption engine to calculate the Integrity Code field, which can save on design costs. However, to handle high-speed data, it may be necessary to have parallel encryption engines to quickly calculate the cipher text and the Integrity Code field.

The authentication operation begins with the generation of an Integrity Code from a nonce using the block cipher in CBC mode. This is followed by the optional

---

[33] It is very difficult, if not impossible, for anyone to guarantee that something is IP-free. The author is not providing any assurances as to the IP status of AES CCM, and implementers need to perform their own analysis.

padded authentication data followed by an optional padded plaintext data. At the receiving side, the same Integrity Code is generated and is then compared to the Integrity Code that is sent with the frame.

Note that the padding only occurs in the generation of the Integrity Code; the actual plaintext is not padded before it is encoded into ciphertext. The key here is that the ciphertext is always the same length as the plain text. If this was not the case, then an additional length field would have been needed in the secure frame formats to indicate the length of the actual plain text.

The selection of the nonce is important to the security characteristics of the symmetric algorithm. The key is to find a nonce that will not have a chance of being repeated before the keys can be changed. The 802.15.3 nonce consists of the SrcID, DestID, Time Token, Secure Frame Counter, and the Fragmentation Control field. The Time Token is 6 octets long and is incremented with every beacon. Even with a 1-ms superframe duration, the time token will roll over only once in every 8,925 years, which should be long enough. However, to ensure that the nonce is unique for every frame sent in a superframe, each secure frame includes a secure frame counter that is incremented for every frame that is sent in the superframe. This allows a DEV to send up to 65,536 frames in a superframe for each DestID before it will need to stop to avoid reusing the nonce. The use of the SrcID allows multiple DEVs in the same piconet to use the same key without running the risk of two frames using the same nonce. The secure frame counter is incremented every time a secure frame is transmitted, even if it is a retransmission.

> There are two types of cryptographic keys defined in the standard: the management key and the data key. While both keys are symmetric keys, the management key is used only to encrypt new data keys for distribution to the DEVs in the security relationship. Each pair of DEVs in a security relationship has a unique set of management keys. On the other hand, within a security relationship, many DEVs might use the same data key. When a DEV leaves a security relationship, the security manager changes the data key using the management key. This prevents the DEV that left from gaining access to the new data key or any data protected by it.

When the Integrity Code for a frame is calculated, the entire 10 octets of the MAC frame header is included in the calculation as is the SECID, secure frame counter, and the payload of the frame. If the frame that is going to be protected is an encrypted data frame, the Integrity Code is computed over the encrypted data, not over the plaintext. If the Integrity Code was calculated over the plain text, then an attacker could instantly determine that a crack had succeeded when the Integrity Code check of the unencrypted data passed. Instead, the Integrity Code only protects information that is already available on the air and so does not give any information regarding the plain text or other secrets.

Key distribution

There are three types of key distribution supported by the standard:

a) The PNC distributing keys to DEVs in the piconet

b) A peer key originator distributing keys to DEVs in the security relationship

c) A DEV requesting the latest key from the key originator

The simplest key distribution protocol is when the PNC needs to deliver new keys to the DEVs in the piconet. This process is illustrated in Figure 6–1.

When a peer key originator distributes a new key to another DEV in the security relationship, it requires a response from the DEV. This response has the effect of being a secure ACK, which is not otherwise provided in the standard. The frame exchange sequence for a peer key originator distributing a key is illustrated in Figure 6–2.

Finally, if a DEV determines that the SECID in the security relationship has changed, the DEV will need to request a new key from the key originator, as illustrated in Figure 6–3. A DEV can determine that the SECID changed for the piconet-wide key by an integrity check failure on the beacon along with a new SECID in the beacon. However, the DEV cannot be sure that the SECID really changed until it requests the new key from the key originator. Because the Integrity Code check failed, the DEV cannot be sure that the PNC actually sent the frame. In a peer-to-peer security relationship, the DEV determines that the SECID has changed when it receives a frame that uses a different SECID and also creates an integrity check failure.

Whenever the SECID changes, the keys change and so the DEV should get Integrity Code check failures whenever there is a new SECID in any of its security relationships.

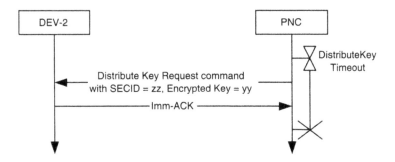

Figure 6–1:   Frame exchange for the PNC distributing a new key to a DEV in the piconet

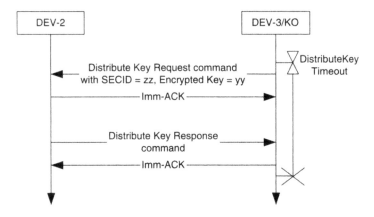

Figure 6–2:   Frame exchange for peer key originators distributing a new key to a DEV in the piconet

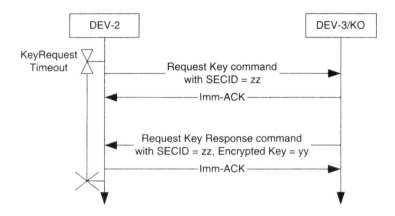

Figure 6–3: **DEV requesting a new key in response to a change in the SECID of the relationship**

SECURITY INFORMATION

The key originator will, in general, maintain an access control list to determine the DEVs that will be allowed to join a secure piconet. When the PNC is the key originator, this also implies determining which DEVs will be able to participate in the piconet as a DEV. Because the ACL includes not only DEVs that are currently in the piconet, but also DEVs that are allowed to join the piconet but are not currently members of the piconet; the list can be quite large. Also, the list is based on the DEV addresses rather than the DEVID because the DEVID can change as DEVs enter and leave the piconet.

When the standard was changed to remove all references to authentication, all references to an ACL were removed as well. Instead, the Security Information Request command and the Security Information Response command were created to facilitate the exchange of security information, in particular to enable DEVs to exchange the ACL. The security information can be retrieved from the key originator either for one DEV at a time or by requesting all of the entries that are maintained by the key originator. To receive a single security information entry from the key originator, a DEV sends the Security Information Request command to the key originator with the Queried DEVID set to the ID of the DEV whose security information is being requested. The key originator will then respond with the security information entry that corresponds to the Queried DEVID. Note that in this case, the DEV is only allowed to request a security information entry for a DEV that is currently part of the piconet. The frame exchange sequence for requesting a single security information entry is illustrated in Figure 6–4.

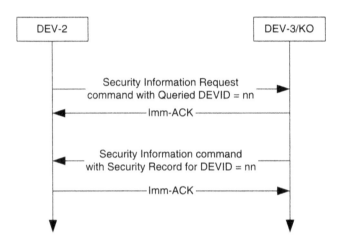

Figure 6–4:   Frame exchange for a DEV requesting a single security information entry from the key originator

When the current PNC hands over control of the piconet to another DEV, it may also want to hand over the ACL. One reason is that the authorization step will generally require the intervention of the user. Requiring this intervention goes against the idea that the hand-over process is seamless and does not interrupt activity in the piconet. The actual authorization protocol may involve bringing the DEVs into close proximity or even touching. If the user had to perform this procedure every time the PNC handed over, no one would want a secure 802.15.3 solution.

The new PNC requests the security information from the current PNC by sending an Security Information Request command to the PNC with the Queried DEVID field set to be the BcstID. The PNC will then respond with a Security Information commands that contain all of the security information records that the

 The information in the Security Information command can be split across multiple commands. However, the command itself cannot be fragmented. Each Security Information command contains a sequence number as well as a field that indicates the total number of frames that will be sent.

PNC is currently managing. The frame exchange sequence for requesting and receiving security information records from the PNC is illustrated in Figure 6–5.

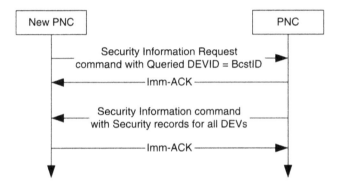

Figure 6–5: Frame exchange for the new PNC requesting all of the security information entries from the PNC

Chapter 7 2.4 GHz PHY

OVERVIEW

The 2.4 GHz PHY uses five modulation formats with an 11 Mbaud symbol rate to achieve scalable data rates from 11 to 55 Mb/s. The modulation types, coding, and resulting data rates are summarized in Table 7–1.

Table 7–1: Modulation, coding, and data rates for 2.4 GHz PHY

Modulation type	Coding	Data rate
QPSK	8-state TCM	11 Mb/s
DQPSK	None	22 Mb/s
16-QAM	8-state TCM	33 Mb/s
32-QAM	8-state TCM	44 Mb/s
64-QAM	8-state TCM	55 Mb/s

The 2.4 GHz PHY was chosen over 5 GHz and ultra-wideband (UWB) proposals for a variety of reasons. The 2.4–2.4835 GHz band is available for unlicensed operation in most of the countries in the world. Although progress is being made in opening up the 5 GHz bands worldwide, there were still many limitations on their use in Europe and Asia. The allocations for UWB have only recently been allowed in the US and are still under consideration in other parts of the world.

Another inherent advantage of using the 2.4 GHz unlicensed band is that it provides a greater range for the same transmit power and receiver design than for 5 GHz radios. Although it is possible to add higher gain antennas to 5 GHz products to extend their range, as is sometimes done with 802.11a products, this results in a decrease in coverage in other areas. For example, some 802.11a access points (APs) use high gain dipole antennas that extend the range of the AP in the horizontal direction. However, this reduces the coverage in the vertical direction, particularly between floors in typical multi-story houses. In the case of UWB, the lower transmit power available as well as the higher frequency range used for UWB (the FCC has allowed intentional emissions for UWB radiators in the 3.1–10.1 GHz range) greatly limits the range that is possible. In particular, at those frequencies and power levels, UWB systems will be limited to use only in a single, small room because of their short range and inability to penetrate walls.

The 2.4 GHz PHY, on the other hand, will easily be able to cover a typical single family home in the US. Using the 802.11b channel model for the 2.4 GHz band, the rate vs. range for an 802.15.3 radio is graphed in Figure 7–1 for both a low-power transmitter (+0 dBm) and a higher power (+14 dBm) one.

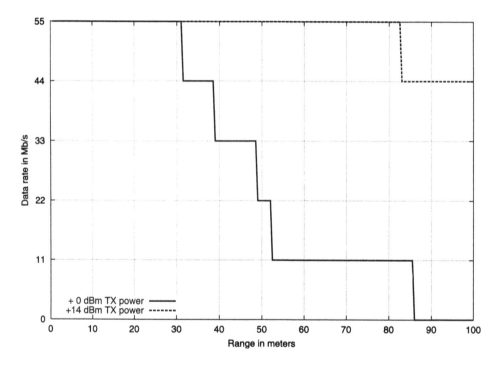

Figure 7–1:   Rate vs. range for high and low power 802.15.3 2.4 GHz PHY

The calculations in Figure 7–1 use the 802.11 channel model (see "Assumptions for coexistence simulations" on page 278), a noise figure of 7 dB, an implementation loss of 4 dB, and 2 dB gain antennas at both the transmitter and the receiver. The noise figure is typical for 802.11b single-chip radios currently on the market. The sensitivity measured at the antenna connector (i.e., excluding the receiver antenna gain) for this configuration is summarized in Table 7–2. A +14 dBm transmit power is typical for 802.11b radios. A power amplifier capable of this type of transmit power uses less than 150 mA at 3 V (for example, the Maxim® 2242 or the SiGe semiconductor PA2423MB). For the 0 dBm case, no power amplifier is required.

Table 7–2: Receiver sensitivity for typical 802.15.3 implementation[a]

Modulation	Data rate	Implementation sensitivity
QPSK-TCM	11 Mb/s	−88 dBm
DQPSK	22 Mb/s	−81 dBm
16-QAM	33 Mb/s	−80 dBm
32-QAM	44 Mb/s	−77 dBm
64-QAM	55 Mb/s	−74 dBm

[a] The standard has relaxed sensitivity requirements, these number represent values for a typical implementation.

Another advantage of a 2.4 GHz radio over 5 GHz or UWB radios is that RF-CMOS and SiGe designs for low-cost, single-chip 2.4 GHz RFICs are available from many sources. The combination of hype for Bluetooth and the brisk sales of 802.11b and 802.11g have encouraged a number of companies to invest in creating these RFICs. In the near future, 5 GHz RFICs will be widely available in low-cost processes from many different sources. Eventually, this will likely be true for UWB ICs, although it will likely take much longer than the ramp-up of 5 GHz technology because high-volume, low-price applications are a brand new market for UWB techniques.

GENERAL PHY REQUIREMENTS

Channel plan

The 2.4 GHz PHY has two channel plans depending on the environment in which it is operating. The center frequencies and channel IDs (CHNL_ID) are listed in

Table 7–3. The high-density channel plan provides four nonoverlapping channels for use when 802.15.3 is the main network in use. If, on the other hand, an 802.11 network is present, particularly if it is occupying 802.11b channel 6 (2.437 GHz center frequency), then the 802.15.3 network will work best with the 802.11b coexistence channel plan. The reason for this is that if the 802.15.3 network used channel 2, it would overlap with an 802.11b network in both channel 1 or channel 6 (802.11b channel numbers). Likewise, if the 802.15.3 network used channel 4, it would interfere with 802.11b networks and be subject to interference from 802.11b networks in both channel 6 and channel 11 (802.11b channel numbers).

Table 7–3:   2.4 GHz PHY channel plan

CHNL_ID	Center frequency	High density	802.11b coexistence
1	2.412 GHz	X	X
2	2.428 GHz	X	
3	2.437 GHz		X
4	2.445 GHz	X	
5	2.461 GHz	X	X

On the other hand, if the 802.15.3 network adopts a channel plan that most closely matches the 802.11b channel plan, it would overlap with at most one channel. If only two of the three 802.11b channels are being used, then the 802.15.3 PNC would search for the best channel, from an interference point of view, and move the piconet to that channel. IEEE Std 802.15.3 requires that if a DEV is capable of detecting an 802.11b network, it needs to adopt the 802.11b coexistence channel plan if it detects an 802.11b network and move the piconet away from channels that contain 802.11b WLANs.

Timing issues

In any radio systems, there are constraints on the timing that arise from the physical limitations of the actual design; e.g., the length of time it takes to change the radio from transmit mode to receive mode. The key timing parameters for the 2.4 GHz PHY are described in Table 7–4. The interframe spacings (IFS's) are used by the MAC in synchronizing with the piconet. The acronyms stand for minimum IFS (MIFS), short IFS (SIFS), backoff IFS (BIFS), and the retransmission IFS (RIFS).

Table 7–4: Timing parameters for the 802.15.3 2.4 GHz PHY

PHY parameter	MAC parameter	Value	Description
Channel switch time	N/A	500 μs	The time required to change physical layer channels.
pPHYMIFSTime	MIFS	2 μs	The time required between successive transmissions.
pPHYSIFSTime	SIFS	10 μs	The time required to switch from transmit to receive and vice versa.
pCCADetectTime	pBackoffSlot	7.3 μs	The time required to determine that the channel is clear.
pPHYSIFSTime + pCCADetectTime	BIFS	17.3 μs	The minimum time required to determine that channel is clear and turn around to begin transmitting.
2*pPHYSIFSTime + pCCADetectTime	RIFS	27.3 μs	The amount of time that needs to expire before the source can send a retransmission in the CTAP.

The channel change timing should be fairly easy to achieve. For example, a Bluetooth radio is required to change frequencies within 224 μs, whereas the channel change requirement for 802.15.3 is twice as large. The only time this number comes into play is when the PNC changes the operating channel of the piconet. The PNC knows that every DEV in the piconet can change to the new channel within 500 μs. Thus, in the last superframe in the old channel, the PNC should not put in any CTAs in the last 500 μs of the superframe so that all of the DEVs have time to switch to the new channel. If the PNC did provide CTAs during this time, the source and destination DEVs would not use that time for communications, but rather would spend the time switching to the new channel. Because channel change is done very infrequently, this time does not affect the overall throughput of the system.

On the other hand, the IFSs occur in essentially every superframe and so they have a direct impact on the efficiency and throughput of the system. The most important parameter for this is the SIFS. This time is used between every CTA, for every Imm-ACK and Dly-ACK frame, in the backoff algorithm, and in the retransmission timing. Therefore, the PHY subcommittee of TG3 wanted to keep this time as short as possible without creating an unreasonable burden on the implementer. Since the 802.15.3 PHY has many characteristics in common with 802.11b, the subcommittee decided to use the same SIFS as 802.11b: 10 µs. The receiver has to be ready before the transmitter begins sending the frame so the TX to RX turnaround is strictly less than the SIFS while the RX to TX turnaround is no less than the SIFS but less than the SIFS plus 1 µs. In the 2.4 GHz PHY description, Clause 11 of the standard, the SIFS time is also referred to as the pPHYSIFSTime.

The MIFS, like the SIFS, has a direct impact on the throughput of the system. Ideally, the radio is able to start transmitting the next frame as soon as it is done with the transmission of the previous frame. In a practical sense, the radio needs a finite amount of time to complete the transmission/reception of one frame before it begins the transmission/reception of the next one. Figure 7–2 shows the percent difference between using a finite time for the MIFS vs. using a zero time for the MIFS as a function of the data rate. The frame has a 2044-octet payload, and the calculation includes the overhead for the PHY preamble, PHY header, MAC header, HCS, FCS, as well as MIFS spacing. Because the MIFS is only used for frames with no-ACK policy, the time for an Imm-ACK or Dly-ACK frame is not included.

In Figure 7–2, the graphs show the amount of throughput lost due to having an non-zero MIFS. Therefore, lower numbers in the graph are better because the goal is to have the minimum amount of overhead. When the MIFS is set at 2 µs, the overhead for the MIFS is less than 2.5% for all data rates. However, for the 10-µs and 20-µs cases, the throughput lost to the MIFS can be as high as 16% at higher data rates. Although the current PHY does not support data rates this high, it is anticipated that a new PHY will be available in 3 to 4 years that could reach rates as high as 400 Mb/s. These reductions in the throughput are in addition to the overhead already in IEEE Std 802.15.3, and so any increase in the overhead is undesirable.

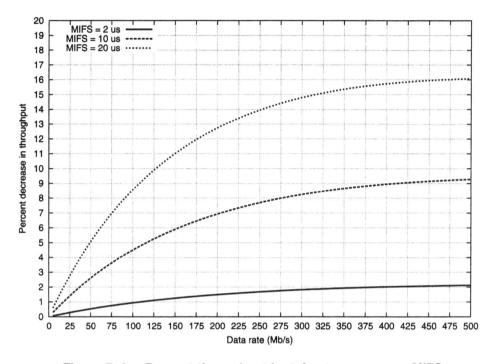

Figure 7–2:   Percent throughput lost due to a non-zero MIFS

The effect of the MIFS value on the throughput of the data rates of the 2.4 GHz PHY is shown in Figure 7–3. The data payload is assumed to be the maximum allowed for this PHY, 2044 octets. For the lowest data rate, 11 Mb/s, the effect of the MIFS on the overall throughput is very low, less than 1% for a 10-µs MIFS and only slightly greater even for a 20-µs MIFS. The throughput decrease increases as the data rate increases. At 55 Mb/s, the difference between using a MIFS of 10 µs vs. 2 µs is about 2.0%, as shown in Figure 7–3. This is an additional 1.1 Mb/s of throughput. Although this improvement is not spectacular, it is important not to discard even small improvements in throughput if they can be implemented in a relatively simple fashion. From the PHY point of view, supporting a 2-µs MIFS is not any more difficult than supporting the TX ramp, which is of the same order. In fact, for a 2-µs MIFS, the transmitter does not get a chance to completely shut off, rather the ramp-down and ramp-up, each of which is 2 µs in duration, overlap in the interframe space. In this case, the input to the transmitter would only be idled between frames and the rest of the TX chain would be left on. The receiver would simply be left on so that it could receive the next frame. As the environment changes slowly relative to the frame duration, keeping the AGC setting from the previous frame would enable fast acquisition of the next frame.

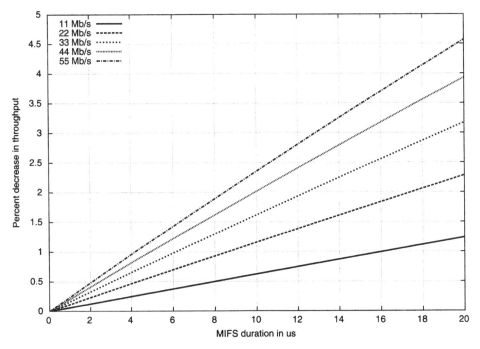

Figure 7–3: Percent throughput lost for each of the supported data rates at various MIFS durations

The MIFS affects the throughput less for larger data frames than for smaller data frames. Figure 7–4 shows the percent difference in throughput at 55 Mb/s due to a change from a 10-μs MIFS, i.e., having the MIFS equal to the SIFS, to a 2-μs MIFS. For larger data payload sizes, the advantage of decreasing the MIFS is diminished. Although the most efficient data transfer in an error-free environment is to use the longest possible frames, this is not true in the error-prone wireless environment. To keep the FER low, the application may use smaller frame sizes because a smaller frame size reduces the FER for a given BER. The improved FER can actually improve the overall throughput even though the overhead is higher for the smaller frames. The application may use smaller data fragments as well, and so it would not be possible for the MAC to use the maximum frame size. For smaller data payloads, e.g., 512 octets, the improvement in throughput is 5.5% or about 3 Mb/s.

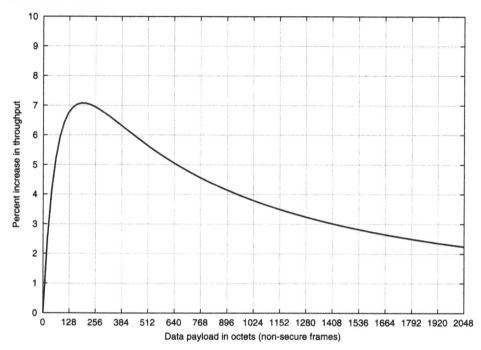

Figure 7–4: Percent throughput increase from changing the MIFS duration from 10 μs to 2 μs for various data payload sizes

Using the CAP for random access requires that the DEVs are able to determine if a frame is being sent over the air. Because the first part of a frame that is being transmitted is the CAZAC sequence, the standard requires that the DEVs can recognize the PHY preamble within five repetitions of the CAZAC sequence. This time, referred to as the pCCADetectTime, is therefore 5*16/11 or about 7.3 μs.

The standard also defines two other interframe spacings, the backoff IFS (BIFS) and the retransmission IFS (RIFS). The BIFS is used in the backoff algorithm to determine if another DEV is transmitting a frame. Once a DEV has determined that it is time for its turn to access the medium in the CAP, it needs to change from receive to transmit and begin transmitting its frame. The other DEVs also need time to hear the start of the frame that is being transmitted. Thus, the BIFS is the sum of the SIFS time and the CCA detection time, i.e., BIFS = pPHYSIFSTime + pCCADetectTime, or about 17.3 μs. The BIFS is also used at the start of the CCA algorithm to ensure that a DEV that successfully sent a frame in the contention period is allowed time to also receive an Imm-ACK, if one was requested.

The RIFS is used for retransmissions in the CTAP. Retransmissions in the CAP are required to use the backoff algorithm. In fact, they are required to use a larger contention window, i.e., the exponential backoff (see "Contention access period (CAP)" on page 55). However, in a CTA (other than a contention-based MCTA), the source DEV is the only one that is allowed to transmit and it can begin its retransmission as soon as it is practical. If the source DEV does not hear an ACK to the frame, it would want to begin to retransmit as soon as possible. In the case of the 2.4 GHz PHY, the source DEV changes from transmit to receive (SIFS) and listens to hear the beginning of the ACK (pCCADetectTime). If it is does not hear the beginning of the ACK, then it changes from receive to transmit (SIFS) and begins to send the retransmission. Thus, the RIFS is equal to 2*SIFS + pCCADetectTime.

There may be two reasons why the source did not get an ACK:

a) The destination did not successfully receive the frame that was sent.

b) The ACK from the destination is being sent, but the source DEV is unable to hear it due to bad channel conditions or because of interference.

In the first case, the most efficient procedure is to send the retransmission as soon as the source DEV determines that the ACK is not going to arrive. In the second case, if the source DEV uses a short RIFS, as defined for the 2.4 GHz PHY, then it will step on the ACK and the receiving DEV will not hear the retransmission either. For this case, the most efficient RIFS would be to choose it to be 2*SIFS plus the duration of the ACK, either Imm-ACK or Dly-ACK. However, the CAZAC sequence has good cross and auto-correlation characteristics, and so it is highly unlikely that the source DEV would be unable to hear any of the CAZACs in the preamble of the ACK and still be able to hear the response. Because of these reasons, the TG decided to use CCA detection instead of waiting for an ACK duration.

Miscellaneous PHY requirements

Throughout the standard, all power measurements are referenced to the antenna connector and do not include the effects of the antenna. It is assumed that the test equipment is matched to the impedance at the antenna connector. If the antenna connector is not available, then the implementer is to estimate the antenna gain and subtract that from the measurement to get a value equivalent to that if the measurement had been made at the antenna-to-radio interface.

The operating temperature range for the measurements specified in this standard is 0 to –40 °C. While most consumer devices will exceed this temperature range, the task group did not want to attempt to determine a single temperature range that would be applicable for all applications. For a device that is used primarily indoors, e.g., a television set, 0 to –40 °C may be sufficient. On the other hand, automotive applications typically require a –40 to +85 °C temperature range.

The MAC protocol allows the PNC to use either the CAP or contention-based MCTAs for DEVs to join the network and to communicate with the PNC. However, this flexibility would require that all DEVs implement both methods in order to communicate with the PNC. The alternative would be that a DEV would be unable to join a piconet because it supported a contention method that was different from the one supported by the PNC. Rather than allow this, the task group decided to require that all PNC-capable DEVs that implement the 2.4 GHz PHY allow the use of the CAP for DEVs to associate and send commands and/or data. The CAP is a very efficient method for channel access (see "Comparing the contention access methods" on page 60) and it benefits low-cost DEVs to implement only one of the two methods. The PNC can also allow DEVs to use MCTAs for contention-based access, but this is in addition to the CAP, not as a replacement for it. As a regular MCTA is exactly the same as a CTA, the PNC is always allowed to use these to communicate with a DEV.

PHY FRAME FORMAT

One of the key requirements for interoperability is to provide a specification for how the data are formatted. Although some of the other requirements involve performance expectations, it is not possible to build interoperable implementations if the devices do not use the same data formats. Thus, the specification of the PHY frame format is one of the key parts of IEEE Std 802.15.3.

Stuff bits and tail symbols

Although the MAC always passes an integer number of octets to the PHY, these bits will not necessarily map into an integer number of symbols. Consider a 3-octet frame payload with a 4-octet FCS. The number of symbols required for this 56-bit MAC frame body are listed in Table 7–5.

Table 7–5: Symbols required to encode a 56-bit MAC frame body

Modulation	Data rate	Bits/symbol	Number of symbols
QPSK-TCM	11	1	56
DQPSK	22	2	28
16 QAM-TCM	33	3	$18\frac{2}{3}$
32 QAM-TCM	44	4	14
64 QAM-TCM	55	5	$11\frac{1}{5}$

Table 7–5 shows that sometimes the MAC frame will require a fractional number of symbols. The MAC frame is always passed as an integer number of octets, and so the 11-, 22-, and 44-Mb/s modes will always be able to encode this as an integer number of symbols. However, the 33- and 55-Mb/s modes are not guaranteed to give an integer number of symbols. For these modes, stuff bits (SBs) are added to the end of the MAC frame body to round out the modulated frame into an integer number of symbols. The transmitting PHY can add either ones or zeros for the stuff bits, and these are ignored by the receiving PHY. The stuff bits are not used as part of the FCS calculation (which technically is done by the MAC). The transmitter always adds just enough bits to complete the symbol. Thus, for the 33-Mb/s mode, it adds less than 3 bits (i.e., 0, 1, or 2), and for the 55-Mb/s mode, it adds less than 5 bits (i.e., 0, 1, 2, 3, or 4).

A typical modem also needs some time to terminate the demodulation process, so extra symbols, called the tail symbols (TS), are added to the end of the frame. For the TCM modes, these tail symbols are specified so that the trellis code always terminates in the same state. Two tail symbols are required for the 16-, 32-, and 64- QAM-TCM modes while the QPSK-TCM mode requires three tail symbols. For the DQPSK mode, two symbols are provided to assist in the implementation of a differentially coherent demodulator. The overhead for the extra symbols is extremely low (<200 ns per frame) while simplifying the modem design by ending the frame in a known set of symbols.

Frame format

The 802.15.3 PHY is responsible for including any PHY-related information in the PHY header and for creating some type of an error check sequence for the combination of the MAC and PHY headers. In addition, for radio synchronization, the PHY prepends a preamble to the frame. The complete PHY frame format for the 22-, 33-, 44-, and 55-Mb/s modes is shown in Figure 7–5.

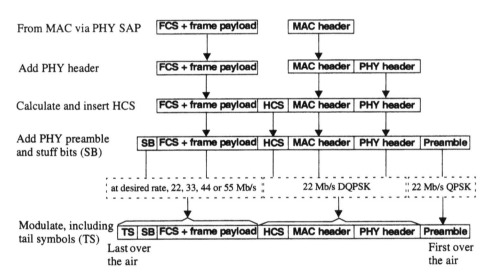

Figure 7–5:   PHY frame formatting for 22-, 33-, 44-, and 55-Mb/s modes

The frame begins with the preamble, which allows the receiver to set its gain control setting, detect and correct for the frequency offset, lock onto the phase of the received signal, and perform any necessary channel equalization to enhance the radio's multipath performance. The modulation for the preamble is 11 Mbaud

QPSK (22 Mb/s). The preamble is followed by the two octet PHY header, which sets the seed for the scrambler, indicates the data rate of the frame payload, and sets the length of the frame payload. The MAC header is sent next, and the PHY calculates a CRC over the PHY and MAC headers and sends this as the two octet header check sequence (HCS). The PHY header, MAC header, and HCS are modulated with the base rate, 22-Mb/s DQPSK. The rest of the frame (if it does not have a zero length frame payload) is modulated at the rate indicated in the PHY header, either 22-, 33-, 44-, or 55-Mb/s. The frame payload and FCS are passed to the PHY from the MAC via the PHY SAP.

The PHY frame for the 11-Mb/s mode is somewhat different. The reason for this is that the 11-Mb/s mode has a much lower required SNR as compared with the 22-Mb/s mode and the error rate for the PHY and MAC headers would determine the overall FER for the 11-Mb/s mode frame. This would completely wipe out the range advantage that is given by using the 11-Mb/s mode. For example, when the signal to noise ratio (SNR) is 5.5 dB, the BER of the 11-Mb/s mode is 10^{-5} while the BER of the 22-Mb/s mode is 2.8×10^{-2}. The FER for a group of bits based on its BER is calculated from

$$FER = 1 - (1 - BER)^{(\text{number of bits})} \tag{1}$$

Thus, the FER for the header and a 2048-byte frame body are

$$FER_{(header)} = 1 - (1 - 2.8 \times 10^{-2})^{(12*8)} = 94.5\% \tag{2}$$

$$FER_{(MAC\ frame)} = 1 - (1 - 10^{-5})^{(1024*8)} = 7.9\% \tag{3}$$

In this case, the overall FER would be 95%. The frame would almost never get through because the header would almost always be in error.

To fix this issue, the PHY header, MAC header, and HCS are sent a second time for 11 Mb/s mode frames, modulated with QPSK-TCM. If the receiver correctly receives the first header, it can ignore the second header. However, if the receiver gets an error in the first header, it will listen for the second header to see if the frame was sent at 11 Mb/s. Because the data rate is in the PHY header, it is possible that the error in the first header sequence is in the bit that indicates the PHY data rate. Thus, if there is an error in the first header, the receiver cannot be sure of the data rate with which the frame is being sent and it must listen for the second header in case the frame is an 11-Mb/s frame.

The PHY frame format for the 11-Mb/s mode is illustrated in Figure 7–6.

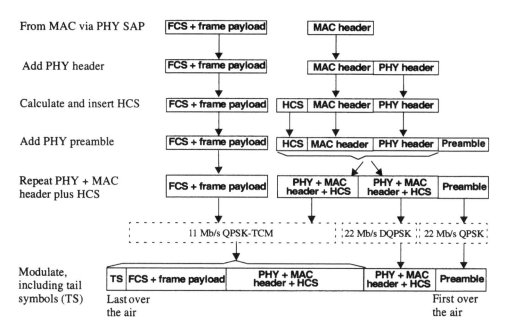

Figure 7–6: PHY frame formatting for 11-Mb/s mode

PHY preamble

In order to synchronize the radio, a known sequence is sent as the first portion of the data frame. For the 2.4 Ghz PHY, this sequence is a constant amplitude, zero-autocorrelation (CAZAC) sequence. For QPSK modulation, the sequence consists of 16 symbols with the phase indicated in Table 7–6. In the table, the phase shift from the previous symbol is also listed, which shows an interesting pattern. The symbols in the sequence consist of four repetitions of each of the four possible phase shifts. As opposed to a PN sequence, the CAZAC sequence has zero autocorrelation when the sequences are not aligned. In a PN sequence, the autocorrelation is generally $1/N$ when the sequences are offset where N is the length of the PN sequence.

Table 7–6: CAZAC sequence

CAZAC sequence element	Value	Degrees phase shift from previous symbol
C_0	$1 + j$	0
C_1	$1 + j$	0
C_2	$1 + j$	0
C_3	$1 + j$	0
C_4	$-1 + j$	90
C_5	$-1 - j$	90
C_6	$1 - j$	90
C_7	$1 + j$	90
C_8	$-1 - j$	180
C_9	$1 + j$	180
C_{10}	$-1 - j$	180
C_{11}	$1 + j$	180
C_{12}	$1 - j$	270
C_{13}	$-1 - j$	270
C_{14}	$-1 + j$	270
C_{15}	$1 + j$	270

The PHY preamble is made up of 12 repetitions of the CAZAC sequences, denoted as $P_0, P_1, P_2, ..., P_{10}, E$. However, the twelfth repetition of the sequence, E, has each element of the CAZAC sequence negated, or equivalently, rotated by 180 degrees. This last sequence acts as a marker to tell the PHY when the first bit of the PHY header will be arriving. The complete physical layer preamble is shown in Figure 7–7.

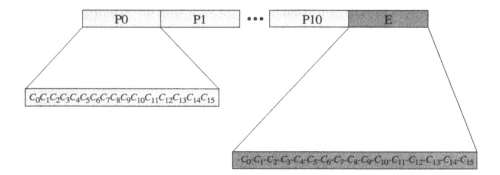

Figure 7–7:   Physical layer preamble format

Data size restrictions

The maximum frame size that is supported by the 2.4 GHz PHY is 2048 octets, including the FCS. Without encryption, this limits the amount of data in a single data frame to be less than or equal to 2044 octets. In addition, the maximum transfer unit size is limited to 2044 octets. This parameter places a restriction on the largest frame that can be given to the MAC for transfer over the PHY. The MAC may fragment the frame into smaller pieces to improve the probability of reception. There are a couple of reasons for this restriction. The first reason is that most protocols actually use smaller frame sizes, e.g., TCP, MPEG, etc. The second reason is that a DEV that is assembling frames as the destination needs to provide sufficient buffering to hold all of the fragments of the frame before it passes them up. In an earlier draft of the standard, the maximum frame size was 65,335 octets, which would require approximately 62 kbytes or storage for each stream that the DEV is receiving. For six streams, this requires a buffer size of over 360 kbytes, which is not consistent with a low-cost implementation.

The MAC does not place a restriction on the smallest fragment. In theory, a DEV could send 1-octet fragments. To prevent this, the smallest fragment allowed for 2.4 GHz PHY implementations is 64 octets. The last fragment may be smaller than this value, and an unfragmented data frame may also be smaller, but the PHY is not allowed to fragment a frame into fragments smaller than this value.

MODULATION

In IEEE Std 802.15.3, five different modulation methods are used to get five data rates with one symbol rate. Figure 7–8 illustrates the signal constellations used in encoding bit streams into discrete signal levels that are sent over the air interface. The base modulation rate, which all DEVs are required to support, is the 22-Mb/s DQPSK uncoded modulation. The base rate is used for all broadcast frames, such as the beacon, because all DEVs are capable of understanding this modulation. The uncoded DQPSK modulation was chosen as the base rate because it is simple to implement and allows the development of compliant radios with a minimum cost. It is even possible to non-coherently demodulate the DQPSK signaling, which would allow extremely simple baseband processing.

However, multimedia applications require high data rates and connectivity over long distances. To achieve these goals, other modulations, up to 64 QAM, with coding in the form of an eight-state trellis code are used. The eight-state trellis encoder is illustrated in Figure 7–9. Note that only the lower two bits for the QAM modulations are used in the actual encoder. The other bits are used to select the symbol from the subsets that are selected by the trellis coder. The set partitioning is described in more detail in the standard. The set partitioning was chosen such that the decimal representation of the bit mapping goes from low to high as the constellation points are traced from center outward. This rule ensures that the decimal representations of the bit mappings from 0 to 15 belong to 16-QAM constellation, and 0 to 31 belong to 32-QAM constellation, and 0 to 63 belong to 64-QAM constellation.

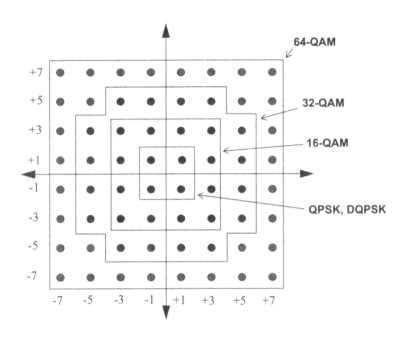

Figure 7–8: DQPSK, QPSK, 16/32/64-QAM signal constellations

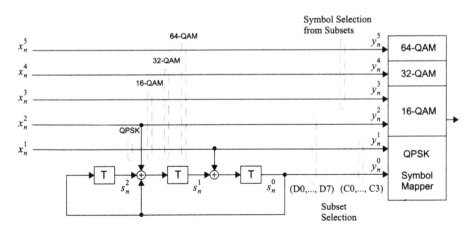

Figure 7–9: QPSK, 16/32/64-QAM eight-state trellis encoder

The final mapping of bits to symbols for the TCM encodings is shown in Figure 7–10. Also shown in the figure are the eight subsets, D0–D7, that are used in creating the mapping from the input bits to the trellis-coded symbols.

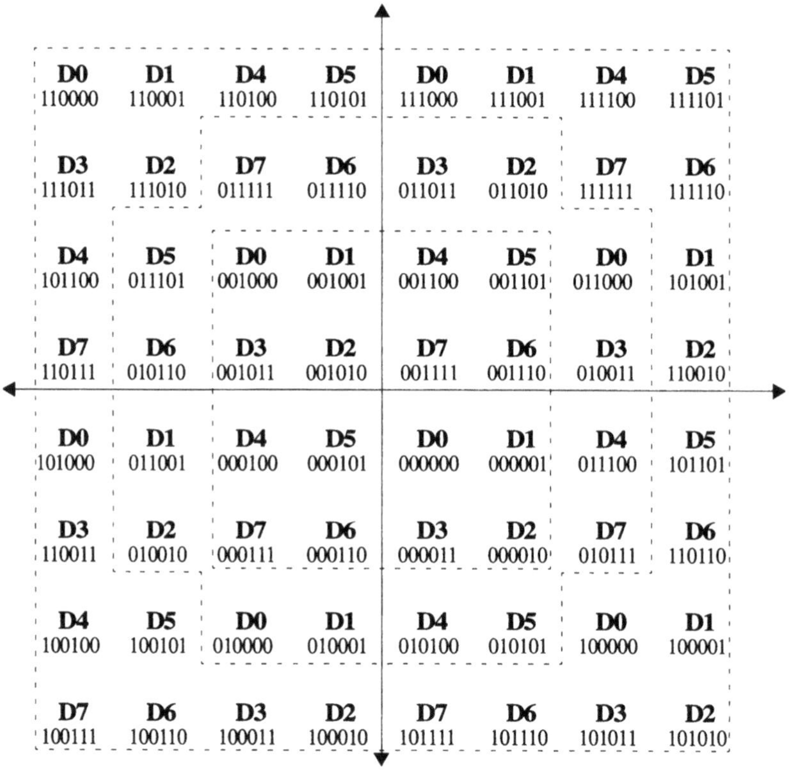

D0 110000	D1 110001	D4 110100	D5 110101	D0 111000	D1 111001	D4 111100	D5 111101
D3 111011	D2 111010	D7 011111	D6 011110	D3 011011	D2 011010	D7 111111	D6 111110
D4 101100	D5 011101	D0 001000	D1 001001	D4 001100	D5 001101	D0 011000	D1 101001
D7 110111	D6 010110	D3 001011	D2 001010	D7 001111	D6 001110	D3 010011	D2 110010
D0 101000	D1 011001	D4 000100	D5 000101	D0 000000	D1 000001	D4 011100	D5 101101
D3 110011	D2 010010	D7 000111	D6 000110	D3 000011	D2 000010	D7 010111	D6 110110
D4 100100	D5 100101	D0 010000	D1 010001	D4 010100	D5 010101	D0 100000	D1 100001
D7 100111	D6 100110	D3 100011	D2 100010	D7 101111	D6 101110	D3 101011	D2 101010

Figure 7–10:   16/32/64-QAM constellation bit mappings

The drawing in Figure 7–8 shows the outer points of the 64-QAM constellation at seven times the distance of the QPSK constellation points. If this was implemented in the radio, the average power in the 64-QAM portion of the frame would be much greater than the power in the preamble, which would cause problems with the receiver's automatic gain control (AGC) algorithms. It would also make the ranges for the 22-Mb/s links closer to the 55-Mb/s links by decreasing the range of the 22-Mb/s mode. To handle the differences in the modulation, the transmitter is required to scale the constellation points such that the average power in the preamble and headers is the same as the average power during the transmission of the data in the frame. This scaling is similar to that used in 802.11a, which also had DQPSK headers followed by up to 16-, 32-, and 64-QAM modulation. The constellation points in Figure 7–8, $I + jQ$, are multiplied by a scaling factor, K_{MOD}, to give the output point, d. The values for K_{MOD} for each of the modulations are listed in Table 7–7.

Table 7–7: Normalization factor for PHY modulation formats

Modulation	K_{MOD}
DQPSK	1
QPSK-TCM	1
16 QAM-TCM	$1/(\sqrt{5})$
32 QAM-TCM	$1/(\sqrt{10})$
64 QAM-TCM	$1/(\sqrt{21})$

The K_{MOD} factor can be calculated by computing the average power in the constellation and setting the average power for DQPSK to be unity. The power of an individual point is the square of the Euclidean distance because the constellation is plotted in voltage. The average power, $P_{avg(mod)}$, then is this number divided by the number of points in the constellation, as detailed below:

$$P_{avg(mod)} = \frac{\displaystyle\sum_{n=0}^{N_{points}-1}(I_n^2 + Q_n^2)}{N_{points}} \qquad (4)$$

The scaling factor is the square root of the ratio of the average power in the DQPSK modulation versus the average power in the other modulation, i.e.,

$$K_{MOD} = \sqrt{\frac{P_{avg(DQPSK)}}{P_{avg(mod)}}} \qquad (5)$$

The values for the total power and average power calculated using this formula for the defined modulations are listed in Table 7–8.

Table 7–8: Total and average power in the constellation

Modulation	Total power	Average power
DQPSK	8	2
16 QAM	160	10
32 QAM	640	20
64 QAM	2688	42

RECEIVER PERFORMANCE

During the development of IEEE Std 802.15.3, there was a consistent focus on keeping the requirements relaxed so that implementers could create low-cost, low-power solutions to best serve the portable, consumer electronics market. This desire to allow for innovative designs from implementers is reflected in the receiver specifications. For example, the standard does not include any intermodulation specifications. DEVs compliant to the standard operate in a highly dynamic environment that varies greatly from user to user. The large disparity between the different use cases made it impossible to generate an interference model that would be applicable to a large number of users. Instead, the intermodulation performance requirements are left to the implementers and users to determine through market forces.

Thus, the receiver and transmitter specifications are focused on guaranteeing a minimum level of performance along with any requirements necessary to guarantee interoperability. The PHY subcommittee defined an error-rate criterion that is used for many of the receiver performance tests. This condition requires the DEV to achieve an FER of 8% with 1024 octet frames. Using Equation (1), this works out to a BER of approximately 10^{-5}. The PHY subcommittee also chose a maximum receiver noise figure of 12 dB and a typical implementation loss of 4 dB. The implementation loss is intended to include imperfections in the analog and digital portions of the receiver. Implementers are free to split this margin in any way they choose. For example, they may use a 14 dB system noise figure and 2 dB of implementation loss or 8 dB of system noise figure and 8 dB of implementation loss. Using these numbers, the minimum required sensitivity for a compliant 802.15.3 receiver is given in Table 7–9. Also listed in the table are numbers for the sensitivity of a typical implementation. It is likely that most implementations will achieve these numbers, although some very low-cost devices may choose values closer to the reference sensitivity.

Table 7–9: Reference sensitivity levels for modulation formats

Modulation	Reference sensitivity	Typical implementation sensitivity
QPSK-TCM	–82 dBm	–88 dBm
DQPSK	–75 dBm	–81 dBm
16-QAM-TCM	–74 dBm	–80 dBm
32-QAM-TCM	–71 dBm	–77 dBm
64-QAM-TCM	–68 dBm	–74 dBm

The SNR required to achieve the error rate criterion is listed for each of the modulations in Table 7–10. Using TCM provides a significant reduction in the SNR number while modestly reducing the data rate. In particular, the 33 Mb/s mode achieves almost the same sensitivity as the DQPSK mode with 50% higher data rate. The TCM encoder is quite simple to realize in the transmitter (see Figure 7–9); however, the decoder is more complex, and a good demodulator requires good design and a non-negligible amount of die area. Because of this, the DQPSK modes are uncoded to allow very simple radios that are able to maintain connectivity with other high-speed 802.15.3 DEVs.

Table 7–10: Required SNR to achieve the error-rate criteria for each of the modulation formats

Modulation	SNR required
QPSK-TCM	5.5 dB
DQPSK	12.6 dB
16-QAM-TCM	13.5 dB
32-QAM-TCM	16.6 dB
64-QAM-TCM	19.8 dB

Figure 7–11 compares the reference sensitivity of 802.15.3 with 802.11b and 802.11a for the data rates supported by those standards. The 802.11g draft currently adopts the reference sensitivities for 802.11a for the mandatory enhanced data rates, so they are the same as the values given in the figure for 802.11a. In IEEE Std 802.11a, the authors included their assumptions—10 dB of noise figure and 5 dB of implementation loss, or 1 dB less than that assumed for the 802.15.3 PHY. The assumptions are not listed in 802.11b or for the original 802.11 DSSS PHY. Taking into account the 1-dB difference in the assumptions of the 802.11a and 802.15.3, the reference sensitivity for 802.15.3 is about 4 dB better than that for 802.11a due mainly to the improved SNR performance of the trellis-coded modulation vs. the combination of the OFDM modulation and convolutional code. The only 802.15.3 modulation that comes close to the 802.11a reference sensitivity is the 22 Mb/s mode, which is essentially the same in the graph but about 1 dB better when the difference in the assumptions are taken into account.

The 802.11b results are somewhat misleading in that the writers of the standard used relatively loose requirements in specifying the sensitivity. For example, the SNR difference between 1 Mb/s Barker coding (1 dB for 10^{-5} BER) and 11 Mb/s CCK coding (7.6 dB for 10^{-5} BER) should have degraded the reference sensitivity by 6.6 dB rather than the 4 dB in the standard. One of the reasons for this may have been that when the initial 802.11 standard was written (1990–1997), low-cost radios were not as sophisticated and so it was more difficult to meet the sensitivity number. By the time that 802.11b was published, 1999, it was much easier to create low-cost radios that also had pretty good performance. In fact, most 802.11b implementations achieve around –90 dBm for the 1 Mb/s mode and about –82 dBm for the 11 Mb/s mode, which brings the values more in line with 802.15.3 reference sensitivity numbers. However, typical 802.15.3 implementations will exceed the reference sensitivity number. Based on the required SNR, an 11 Mb/s 802.15.3 radio should have about 2 dB better performance than an 802.11b radio at the same data rate.

One of the other important parameters for a radio is its ability to reject jamming signals. In IEEE Std 802.15.3, as in other standards, this performance is simulated by introducing another instance of the signal, uncorrelated in time with the desired signal, in one of the other channels and determining the error rate performance in the desired channel. Because the jammers in the 2.4 GHz band can be quite

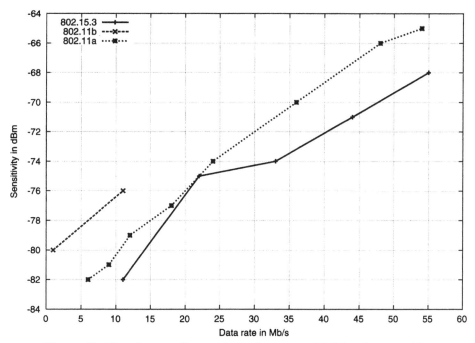

Figure 7–11: Comparison of receiver sensitivities for 802.11a, 802.11b, and 802.15.3

varied, e.g., 802.11 FHSS, 802.11b, 802.15.1, cordless phones, microwave ovens, and so on, it would be much too difficult to both specify performance criteria for each of these jammers as well as for implementers to test to all of these. Instead, an 802.15.3-compliant transmitter is introduced in an adjacent channel with random DQPSK data uncorrelated in time with desired signal. The desired signal is set at 6 dB above the reference sensitivity to remove the effects of thermal noise. The jamming signal is then put into one of the other channels, either adjacent or alternate, with a power level equal to the reference sensitivity plus the value indicated in Table 7–11.

For example, for 16 QAM-TCM, the desired signal is set at –68 dBm (–74 dBm + 6 dBm) while the interfering signal is set to either –49 dBm in the adjacent channel or to –34 dBm in the alternate channel. A summary of the values for the desired signal and the jamming signal power is given in Table 7–12.

For the purposes of the test, the four-channel plan is used (i.e., channels 1, 2, 4, and 5) and an adjacent channel is defined as a channel that is one removed from the desired channel while the alternate channel is at least one channel further out. A list of adjacent and alternate channels for the jamming resistance tests is given in Table 7–13.

Table 7–11: Receiver jamming resistance requirements

Modulation format	Adjacent channel rejection	Alternate channel rejection
QPSK-TCM	33 dB	48 dB
DQPSK	26 dB	41 dB
16 QAM-TCM	25 dB	40 dB
32 QAM-TCM	22 dB	37 dB
64 QAM-TCM	19 dB	34 dB

Table 7–12: Power levels for desired and jamming signals

Modulation format	Desired signal	Adjacent channel power	Alternate channel power
QPSK-TCM	–76 dBm	–49 dBm	–34 dBm
DQPSK	–69 dBm	–49 dBm	–34 dBm
16 QAM-TCM	–68 dBm	–49 dBm	–34 dBm
32 QAM-TCM	–65 dBm	–49 dBm	–34 dBm
64 QAM-TCM	–62 dBm	–49 dBm	–34 dBm

TRANSMITTER PERFORMANCE

The transmitted signal is band limited to allow up to four channels in the 83.5 MHz allocated for unlicensed operation in the 2.4 GHz band. Although the FCC has allocated spectrum from 2.4 GHz–2.4835 GHz, there are "restricted bands" at 2.31–2.39 GHz and at 2.4835–2.5 GHz. The FCC restricts the field strength emitted by an intentional radiator in these bands. The allowed signal level in the restricted band is approximately –42 dBm equivalent isotropically radiated power (EIRP) in a 1-MHz bandwidth. EIRP means the power input to an ideal

Table 7–13: Adjacent and alternate channels for receiver jamming resistance test

Desired channel number	Adjacent channel number	Alternate channel number
1	2	4, 5
2	1, 4	5
4	2, 5	1
5	4	1, 2

isotropic radiator that results in the same field strength. As 802.15.3 devices will typically transmit about +14 dBm in a 15-MHz bandwidth, the in-band power is about +3 dBm/MHz. To achieve the FCC requirements, the transmitter will need to ensure that the transmit signal power in the restricted band is down by approximately 45 dB. The closest restricted band is at the lower band edge, separated by 22 MHz from the center frequency of channel 1. The higher frequency restricted band is 21.5 MHz from the center frequency of channel 5.

A compliant 802.15.3 2.4 GHz PHY transmitter is required to meet the power spectral density limits specified in Table 7–14. These limits were developed to balance the desire for piconets that overlapped in space on separate channels while keeping the transmitter complexity low. Decreased limits (i.e., lower allowed power) improve the performance of 802.15.3 networks that are located near each other in space but are on different channels. However, as this limit is decreased, the linearity requirements for the PA are increased. If the limits are too low, special filtering and even different transmitter architectures would be required to achieve the required performance.

Table 7–14: Transmit PSD limits

Frequency	Relative limit
$7.5 \text{ MHz} < |f-f_c| < 15 \text{MHz}$	-30 dBr
$15 \text{ MHz} < |f-f_c| < 22 \text{MHz}$	$-10/7[|f-f_c \text{ (MHz)}| + 13]$ dBr
$22 \text{ MHz} < |f-f_c|$	-50 dBr

A graphical representation of the transmit PSD is shown in Figure 7–12. The transmit spectral requirements for 802.15.3 are more stringent than either 802.11b or 802.11a. There are two reasons for this: the first is to allow up to four channels in the 73 MHz of usable bandwidth and the second is to improve the coexistence properties of 802.15.3 with other wireless standards.

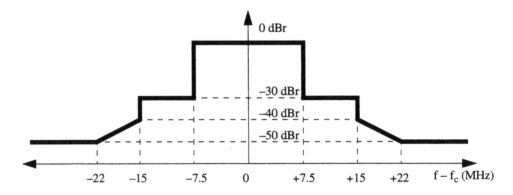

Figure 7–12:   Transmit power spectral density mask

The maximum transmitter power is not set by IEEE Std 802.15.3. Rather, implementers are free to use any power up to the limits allowed in the geographical region where they want their products to operate. A summary of the requirements in four of the geographic regions are listed in Table 7–15. As indicated in the table, there are two rules that are applicable to 2.4 GHz PHY in the U.S.: 15.247 and 15.249 [B1]. The 15.247 rules originally required that the device use a form of spread spectrum, either frequency hopping or direct sequence. Direct sequence spread spectrum (DSSS) systems had to pass a jamming resistance test to be certified under this ruling. However, in early 2002, the rules were modified to change from requiring a DSSS system to only requiring digital modulation. In addition, the jamming margin test was replaced by a power spectral density limit to ensure that the power was spread over a reasonable bandwidth. The higher power allowed under the new ruling allows 802.15.3 systems to achieve data rates and ranges in excess of those offered by either 802.11b or 802.11a.

Table 7-15: Maximum transmit power levels

Geographical Region	Power limit	Regulatory document
Japan	10 mW	ARIB STD-T66 [B2]
Europe (except Spain and France)	100 mW EIRP 10 mW/MHz peak power density	ETSI EN 300 328 [B6], ETSI ETS 300 826 [B7]
USA	50 mV/m at 3 m in at least a 1 MHz resolution bandwidth[a]	47 CFR 15.249 [B1]
USA	1 W total power and less than 8 dBm in any 3 kHz bandwidth, reductions for antenna gains > 6 dB	47 CFR 15.247 [B1]

[a] Electric field strength measurement rather than conducted power measurement.

IEEE Std 802.15.3 also places a requirement on the signal fidelity of the transmitted signal. This requirement is measured by comparing the actual transmitted constellation to the ideal constellation. This measurement, called the error vector magnitude (EVM), is illustrated in Figure 7-13. The EVM performance is very important because a high EVM will make it much more difficult for the receiver to successfully demodulate the transmitted signal.

The calculation for the EVM is defined as:

$$EVM = \sqrt{\frac{\frac{1}{N}\sum_{j=1}^{N}\left(\delta I_j^2 + \delta Q_j^2\right)}{S_{max}^2}} \times 100\% \tag{6}$$

where S_{max} is the magnitude of the vector to the outer-most constellation point and $(\delta I_j, \delta Q_j)$ is the error vector as shown in Figure 7-13. The allowed EVM values for 802.15.3 DEVs are listed in Table 7-16.

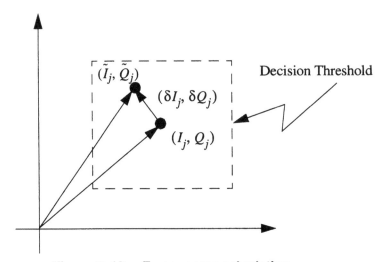

Figure 7–13: Error vector calculation

Table 7–16: EVM values for various modulations

Modulation	r (dB)	SNR (dB)	EVM (%)
64 QAM	3.6796	26	3.3
32 QAM	3.3606	23	4.8
16 QAM	2.5527	20	7.5
DQPSK/QPSK-TCM	1.7700	19	9.2
QPSK-TCM	1.7700	12	20.0

The EVM values shown in Table 7–16 have been calculated using the indicated SNR values in the table. These SNR values incorporate the TCM coding gain (except for QPSK). The values for the SNR in the table are at least 6 dB greater than the SNR required to achieve a 10^{-5} BER. The equation used to calculate the EVM comes directly from vector considerations of the signal plus noise in relationship to the ideal constellation points of any M-QAM signal. This equation for calculating the EVM is:

$$SNR_{M\text{-}QAM} = -(r + 20 \log [EVM_M /(100\%)]) \qquad (7)$$

where *r* is the peak-to-average energy ratio. The value for 32-QAM is an approximation because the constellation is a "cross;" thus, some penalty is incorporated due to an increased encoding and decoding complexity among the constellation points; i.e., not all constellation points are separated by the same Hamming distance. The minus sign in Equation (7) is necessary because SNR is the ratio of signal to noise, whereas EVM is the ratio of noise to signal.

REGULATORY AND REQUIREMENTS

The 802.15.3 2.4 GHz PHY was designed such that it could be used essentially worldwide. However, IEEE Std 802.15.3 does not specify the performance required to achieve regulatory compliance in the various geographic regions where this band is allocated for unlicensed use. For example, the requirements for spurious emissions are not listed in the standard; instead, the implementer is required to adhere to the regulations appropriate for the regions in which they wish to sell their products.

The list of the geographical regions and the appropriate regulatory agencies are listed in Table 7–17.

Table 7–17: Geographic regions and regulatory bodies

Region	Approval standard	Document	Approval authority
Europe[a]	European Telecommunications Standards Institute (ETSI)	ETS 300-328, ETS 300-826	National type approval authorities
Japan	Association of Radio Industries and Businesses (ARIB)	ARIB STD-T66	Ministry of Post and Telecommunications (MPT)
United States	Federal Communications Commission (FCC)	47 CFR, Part 15, Sections 15.205, 15.209, 15.247, 15.249	FCC
Canada	Industry Canada (IC)	RSS-210 [B25]	IC

[a]   Except France and Spain.

Many of the regulatory documents are available online as well. Table 7–18 lists the location of the documents for the various approval authorities.

Table 7–18: Online document locations for regulatory agencies

Organization	Documents available from
ETSI	http://www.etsi.org
ARIB	http://www.arib.or.jp/[a]
FCC	http://www.fcc.gov/oet/info/rules/Welcome.html
IC	http://strategis.ic.ga.ca/engdoc/main.html

[a] ARIB documents can be difficult to obtain.

DELAY SPREAD PERFORMANCE

One of the PHY characteristics that was evaluated in the downselection process was the ability of the proposed PHY to withstand delay spread. Delay spread, or multipath, is the result of reflections of the transmitted signal from objects in the environment that reach the receiver. As an example, consider only two paths, the direct path to the receiver and a reflected path, as shown in Figure 7–14. In the figure, the straight line represents the direct ray or the line of sight (LOS) signal. The other two paths, d_1 and d_2, combine to provide the indirect ray or the non-line of sight (NLOS) signal. Because the NLOS signal travels farther and will generally lose energy in the reflection, it is usually a lower amplitude than the LOS signal.

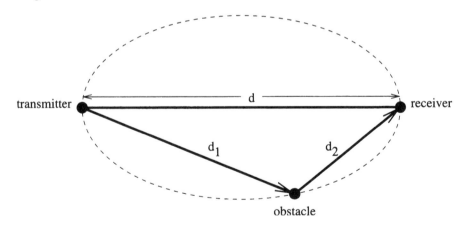

Figure 7–14: Geometry for two-ray multipath

If the path difference between the two rays is an odd multiple of one-half the wavelength, then the amplitude of the signal from the direct path will be reduced by the amplitude of the reflected wave. If the direct and reflected signals are nearly the same in amplitude, the cancellation will be nearly complete. The fading from this effect can be as much as 30 or 40 dB in a narrow bandwidth. Generally, the deeper the fade, the more narrow it will be in both frequency and position. For a 2.45 GHz frequency, one-half of the wavelength in free space is 61 mm or about 2.4 inches. The result of this is that typically, there are many deep nulls (greater than 20 dB) in the 2.4–2.4835 GHz frequency range for typical environments. Because both the radios and the objects around them may be moving, the frequency of these nulls will change as a function of time.

If, on the other hand, the time difference between the paths of the two rays is on the order of a symbol period, then the received signal will be distorted because symbol n will overlap with symbol $(n + 1)$, increasing the BER of the channel through intersymbol interference (ISI). A rule of thumb for the limit on this time difference is that it should be less than 15% to 25% of the symbol period. The limits on the delay spread, using this rough criteria, for 1 Mbaud (Msymbol/s) and 11-Mbaud systems, are summarized in Table 7–19.

Table 7–19: Delay spread limits for two symbol rates

Symbol rate	Symbol duration	15% time	15% distance	25% time	25% distance
1 Mbaud	1000 ns	150 ns	45 m	250 ns	75 m
11 Mbaud	91 ns	14 ns	4.1 m	23 ns	6.8 m

The actual delay spread that can be tolerated also depends on the packet length, type of modulation, protocol, desired packet error rate, type of receiver, and the actual distribution, in amplitude and time, of the delayed signals. In addition, because the LOS signal is usually much stronger than the NLOS signal, the effect of delay spread will be worse if there is not a direct ray from the transmitter to the receiver.

Mitigating the effects of delay spread

Unlike most noise sources in radio systems, increasing the transmitted signal power does not improve the BER once delay spread has become the dominant error term. The reason for this is that the interfering signal is generated by the transmitter, and so increasing the transmit power also increases the interfering power. Decreasing the noise figure of the system also has no effect because the thermal noise power is dwarfed by the power of the interferer.

There are techniques for minimizing the effects of delay spread. These include the following:

* Equalization, both fixed and adaptive (e.g., Viterbi, decision feedback equalizer, linear equalizer, etc.)
* Spreading the signal bandwidth, e.g., DSSS and OFDM
* Coding techniques, such as FEC and interleaving
* Directivity by using higher gain antennas
* Diversity, either space, polarization, frequency, or a combination of these

Coding modifies the signal to give it properties (such as good auto-correlation) that are resistant to fading. This approach consumes more bandwidth in order to reduce fading. Equalization, on the other hand, sums the received signal with copies of the signal that are delayed in time and are adjusted in amplitude to undo the effects of multipath. Equalization methods must be adaptive to handle the dynamic nature of the channel. Equalization techniques can be very effective, but they add complexity and cost to the system. Directivity is an simple solution, but it requires that the transmitter and receiver are "pointed" at each other and it results in larger antennas. Diversity uses multiple sources for the received signal, choosing the "best" source to make the decision. Diversity increases the complexity, cost, and size of the system. Spatial diversity with antenna patterns that are nearly orthogonal in azimuth and elevation allow higher directivity without giving up near omnidirectional coverage.

Fading channel model used for 802.15.3

To model the effects of multipath, an exponentially decaying Rayleigh fading channel was chosen as the environmental model to compare the proposals. The model was originally proposed in IEEE P802.11-97/96[34]. The channel is assumed to be static throughout the frame and is generated independently for each frame. An example of the channel impulse response using this model is illustrated in Figure 7–15. In the figure, the black arrows illustrate the average magnitude of the signal over time. The gray arrows illustrate the magnitude of a specific random realization of the channel. The time positions of black and gray samples are staggered for clarity only. In the model, the channel has values only at the sampling instants, i.e., at $t = T_s$.

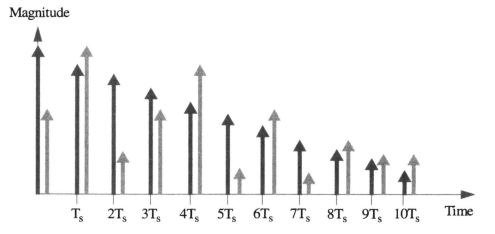

Figure 7–15:   Channel impulse response used for 802.15.3 evaluation

[34] This is an 802.11 Working Group document number which indicates that this was submitted in 1997 and was the 96th document that year. The document can be located via the 802.11 Working Group web page, http://grouper.ieee.org/groups/802/11/Documents/DocumentArchives/1997_docs in the file 97sep.zip with the file name 70967.doc.

The impulse response of the channel, h_i, is composed of complex samples with random uniformly distributed phase and Rayleigh distributed magnitude with average power decaying exponentially, as shown in Figure 7–15. The impulse response is then given by

$$h_i = N\left(0, \tfrac{1}{2}\sigma_k^2\right) + jN\left(0, \tfrac{1}{2}\sigma_k^2\right) \tag{8}$$

$$\sigma_k^2 = \sigma_0^2 e^{-k(T_s/T_{RMS})} \tag{9}$$

$$\sigma_0^2 = e^{-T_s/T_{RMS}} \tag{10}$$

where

$$N\left(0, \tfrac{1}{2}\sigma_k^2\right) \tag{11}$$

is a zero mean Gaussian random variable with variance

$$\tfrac{1}{2}\sigma_k^2, \tag{12}$$

and σ_0^2 is chosen so that the condition

$$\sum \sigma_k^2 = 1 \tag{13}$$

is satisfied to ensure the same average received power.

The sampling time, T_s, in the simulation is assumed to be shorter than a symbol time (or chip time) by at least a factor of four. In simulations T_s is typically a submultiple of the symbol duration. The number of samples to be taken in the impulse response should ensure sufficient decay of the impulse response tail, e.g., $k_{max} = 10T_{RMS}/T_s$.

Defining delay spread performance

Although it is common to quote delay spread performance in terms of time, e.g., "Multipath delay spread tolerance > 250 ns,"[35] that single number does not even begin to describe the delay spread performance of a radio system. The channel model is one of the key items that needs to be fully specified in order to determine the delay spread performance of a system. Different architectures and modulations can have markedly different performance characteristics depending on the channel model that is chosen. The channel model will either have to specify the exact number and type of rays (distance and amplitude) in the model or it will have to assume a distribution (typically either Rician or Rayleigh) and specify the rms delay spread as well as the standard deviation of the delay spread. In addition to that, the introduction of delay spread will increase the FER of the system so another key aspect of the specification is the acceptable level of degradation of the system FER due to delay spread. Finally, the received signal level needs to be specified.

The 802.15.3 PHY subcommittee formed a subgroup to draft a concise definition of delay spread so that all of the proposals could be evaluated against the same criteria. This definition and the results of the evaluations are contained in the IEEE TG3 document 00/110r14P802-15_TG3-Criteria-Definitions.doc. The test required the following steps:

a) Generate 1000 random channels according to environment model, for the desired delay spread T_{RMS}.

b) Apply a signal gain, such that the average RF signal level of the 1000 random channels is 14 dB higher than the RF signal level required to achieve a 1% FER in an AWGN channel. Because the channel model is Rayleigh faded, each instance of the channel may be either above or below the average power.

c) Simulate the FER for each of the 1000 random channels. The simulation should include AWGN arising from the RF circuit (thermal noise + noise figure). Perfect symbol and carrier synchronization may be assumed. Equalizers must perform frame-based coefficient adaptation. The

---

[35] From Texas Instruments' product bulletin on the ACX100 802.11b baseband processor, http://focus.ti.com/pdfs/vf/bband/ti_acx100.pdf.

simulation should not include advanced channel selection and/or diversity techniques that alter the distribution of the channel model. Use one of the two simulation methods defined below:

1) *Direct measurement of FER:* Simulate at least 1000 frames per random channel. Each frame should consist of the proposed preamble, header, tail, and 512 bytes of data. Directly calculate the FER for each random channel.

2) *Measurement of FER by BER:* Simulate the BER for each random channel. The measurement may be performed on continuously transmitted data; however, equalizer adaptation must be performed on the proposed preamble/header. Convert the BER to an FER, assuming 512 bytes of data, preamble, header, and tail bits.

d) Discard the results of the 50 channels (5%) with the highest FER. Find the maximum FER of the remaining channels.

e) Repeat steps a) through d) for different values of T_{RMS}. $T_{RMS,}$ values of 10, 25, and 40 ns must be simulated. Additional values of T_{RMS} may also be simulated to demonstrate the robustness of the system: 50, 75, 100 ns, and so on. The delay spread tolerance is the maximum value of T_{RMS} in which the maximum FER over 95% of the channels is at most 1%. All lower values of T_{RMS} must also achieve 1% FER.

Delay spread measurements

With the channel model and simulation procedure, the only element left was to determine an appropriate value of T_{RMS} that reflected typical values in the home environment. A review of the literature indicated that a value of about 25 ns was appropriate. Huang and Khayata [B9] performed delay spread measurements at 915 MHz, 2.44 GHz, and 5.8 GHz with dipole antennas. The areas that they measured included office space (with cubicles), library, auditorium, classrooms, and a basement with large motors and electrical equipment. Their observation of the delay spread was as follows (note, F1 is 915 MHz):

> It can be observed from Fig. 1 that delay spreads are under 100 ns in most of the cases. The median level of delay spreads is under 50 ns is most of the cases. Delay spreads do not change much as a function of

frequency except in the Auditorium area. We found, by examining individual multipath delay profiles, that radio wave tends to be reflected more by building walls and other internal structures at F1 than at the two higher frequencies. ... At all three frequencies (more observed at higher frequencies), the regression curves show that delay spreads increase slowly with respect to distance.

They also noted that, "The linear regression graphs also shows that the delay spread at 5.8 GHz lower than that at 2.44 GHz for short distances. In increases more rapidly, however, and appears to overtake between 128 and 265 feet." These measurements give a value of less than 50 ns for environments, e.g., office areas, that should have much worse delay spread than a typical home. John C. Stein in an article on the Harris (now Conexant®) web page [B26] indicated that in an office environment, "Typical values for indoor spreading are less than 100 nanoseconds," which is similar to the results obtained by Huang and Khataya.

Nobles et al. [B23] measured delay spread in an office building at three frequencies. They used swept frequency measurements with an IFFT to calculate the time domain data. They found the following (note that [2–4] below are references in the paper):

> The values of the mean rms delay spread (10–50 ns) are typical of those found in similar indoor environments [2–4]. For location T1 the LOS values are very similar to those for location T2 despite the quite different placing of the transmitter within the Communications Lab. In both cases mean values of rms delay spread reduce with increasing frequency.

The authors provided both the mean and the standard deviation of the measured delay spread at three different locations for both LOS and NLOS configurations. The results from the measurements are summarized in Table 7–20 (in the table, the columns labeled s.d. are for the standard deviation). These measurements support a more benign view of the delay spread environment. The measurements are somewhat unique in two ways. The first is that the authors measured three frequencies that are dramatically different. The second difference is that the authors used a swept frequency measurement rather than a pulse response type measurement.

One of the best known researchers in the area of channel measurements is Dr. Theodore S. Rappaport, now a professor at the University of Texas at Austin. In a 1992 paper, Rappaport and Hawbaker [B24] reported on measurements taken at five sites: a typical sports arena (Site A), two (dissimilar) factory buildings (Sites B and C), a six-story closed-plan office (Site D), and a large single-story grocery store (Site E). The measurements were taken at 1.3 and 4 GHz, and they provided cumulative distribution functions (CDF's) of the delay spread. The authors summarize the results for the delay spread measurements by stating the following (note: σ_{tau} is the rms delay spread and [1] is a reference to the thesis of Hawbaker [B8]):

> Most σ_{tau} values above 85 ns were measured in Site B, the large open-plan factory that contains metal machinery and inventory. Most values of σ_{tau} below 50 ns are for the closed-plan office building (Site D) which contained partitioned walls and standard dropped-ceilings. Median values of σ_{tau} range from as low as 22 ns for Site D to 67 ns for Site B. Extensive delay spread results are given in [1], and show very little difference between 1.3 GHz and 4.0 GHz, or between high and low base station antenna heights.

Table 7–20: Measured mean and standard deviation of delay spread in three locations from Nobles, et al. [B23]

Location		Frequency					
		2–2.5 GHz		5–6.5 GHz		17–17.5 GHz	
		mean	s.d.	mean	s. d.	mean	s. d.
T1	LOS	34.5	13.2	14.4	3.0	11.0	2.2
	NLOS	39.6	10.5	22.0	4.5	15.3	8.4
T2	LOS	37.5	11.8	14.8	3.0	11.6	1.6
	NLOS	37.7	12.0	21.2	9.4	19.2	10.5
T3	LOS	49.0	5.4	15.7	11.7	26.9	17.7
	NLOS	37.6	17.0	22.1	5.5	27.6	10.6

The rms delay spread measured in the paper is summarized in Table 7–21. Four antenna configurations were used, and measurements were made at different heights. These measurements characterize the channel in an office and industrial environment. These types of environments tend to have larger rms delay spread due to the high reflectivity of equipment and cubicles (which have metal mesh inside).

Table 7–21: Delay spread measurements from Rappaport and Hawbaker [B24]

Configuration and elevation	σ_{tau} = rms delay spread (ns)			
	Directional LP[a]	Directional CP[b]	Omni LP Vertical	Omni LP Horizontal
LOS - 9.40 m	5.1	1.9	14.0	12.9
LOS - 15.2 m	13.0	2.9	34.4	28.7
LOS - 31.1 m	4.3	2.1	33.6	29.2
NLOS - 34.1 m	33.9	46.1	57.8	61.4
NLOS - 31.7 m	25.6	44.2	79.9	66.9
NLOS - 13.6 m	34.2	26.3	55.1	62.7

[a] LP – A linearly polarized (LP) antenna
[b] CP – A circularly polarized (CP) antenna)

In a paper by MacLellan et al. [B20], the delay spread and path loss in four woodframe residential homes were measured in the 915 MHz and 1900 MHz bands. For the path loss measurements, the authors determined an effective path loss exponent and the standard deviation for the homes and for the average of all of homes. In terms of delay spread, the authors state:

> The multipath measurements suggest that the time dispersion in both LOS and NLOS topographies may be slightly less at 1900 MHz than at 915 MHz. Typical rms delay spreads for LOS topographics were 9.76 ns for 915 MHz and 8.26 ns for 1900 MHz. Similarly, average rms delays for all NLOS measurements were 14.39 ns for 915 MHz and 12.47 ns for 1900 MHz.

The average standard deviation of rms delay spread from the paper is listed in Table 7–22.

Table 7–22: Standard deviation of delay spread measurements from MacLellan et al. [B20]

Condition	915 MHz	1900 MHz
LOS: Indoor-to-Indoor	2.89	2.42
LOS: Outdoor-to-Indoor	1.71	3.77
NLOS: Indoor-to-Indoor	6.51	4.61
NLOS: Outdoor-to-Indoor	2.62	2.87

Very recently, the Intel® Corporation contracted with the University of Southern California to perform extensive channel measurements in a residential environment. The database of the measurements is available on line at the Intel UWB Database [B19] and contains nearly 900 residential channel measurements that span a frequency band of 2–8 GHz with a resolution of 3.75 MHz. The website allows the user to request delay spread and path loss data for a given frequency range. Using a frequency range of 2.4 GHz to 2.4835 GHz, the results for LOS and NLOS measurements are presented in Table 7–23. The database returns three types of calculations, frequency domain, time domain and windowed time domain with RMS values for all three and mean values for the latter two. The largest value of either the RMS or mean delay spread is less than 20 ns for both LOS and NLOS measurements.

Based on the measured data presented in these papers, a 25 ns rms delay spread for the home environment is a reasonable number.

RADIO ARCHITECTURES

The 802.15.3 PHY was developed to enable low-cost implementations. However, in order to realize these cost savings, the radio designer needs to choose an architecture that meets the requirements of 802.15.3 and is cheap to build. The choice of a radio architecture will change the impact of these requirements on the design of the radio.

Table 7–23:   Average delay spread measurements from Intel UWB Database [B19]

Calculation method	LOS	NLOS
Mean τ_{RMS} (frequency domain approach)	18.03 ns	18.48 ns
Mean τ_{RMS} (time domain approach)	10.21 ns	13.53 ns
Mean τ_{RMS} (windowed time domain approach)	7.51 ns	11.93 ns
Mean τ_{mean} (time domain approach)	10.45 ns	16.30 ns
Mean τ_{mean} (windowed time domain approach)	6.97 ns	12.92 ns

The discussion in this section centers on the architectures for the receiver. In most cases, the transmitter architecture will be similar. The transmit-to-receive turn-around time is very short, which requires that the TX and RX frequency sources be locked simultaneously. Other than introducing an extra set of synthesized sources, and thereby increasing the cost of the transceiver, the receiver VCOs will likely need to be reused for the transmit chain. Thus, the frequency plan for the transmitter would likely be the same as for the receiver.

Superheterodyne

The 802.15.3 standard's 2.4 GHz PHY was based on many of the characteristics of 802.11b. Current (as of 2002) 802.11b radios on the market use a superheterodyne architecture with an IF in the 350–450 MHz range, as shown in Figure 7–16. This architecture allows for excellent channel filtering and eases the problem of I/Q phase and amplitude imbalance. One reason for this is that the I/Q split is accomplished at a lower frequency. One method for generating good quadrature out of the second LO is to run the VCO at four times the frequency of the second LO and use a divide by four circuit. A divide by four allows near-perfect quadrature because the phase difference of the LO is independent of the duty cycle of the VCO. In the case of a divide by two, one of the phases is derived from the rising edge and the other from the falling edge. If the VCO has a perfect 50% duty cycle, then this will give accurate quadrature phase. In the case of the divide by four, both phases can be derived from either the rising or falling edges of the VCO, and so the result is independent of the duty cycle of the VCO. With a 350–450 MHz IF, the VCO frequency would be in the range of 1.4–1.8 GHz. This will make the

VCO smaller and potentially easier to realize on-chip due to the reduced inductor and varactor size required for this frequency.

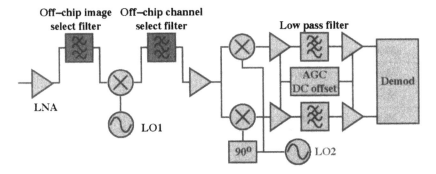

Figure 7–16: Superheterodyne radio architecture

However, the superheterodyne architecture does require a channel select filter with a fairly narrow bandwidth (<5%). A narrow bandpass filter requires high-Q components, which leads to the requirement for a surface acoustic wave (SAW) filter. In the past, SAW filters have been large and expensive. The SAW industry has been able to reduce the size somewhat and has made great advances in reducing the price. In fact, a SAW filter for an 802.11b application is now under $1.00[36] in large volumes. Although the size of the filter has decreased, it is still almost as large as the package for a direct conversion radio IC. In some applications, e.g., an 802.15.3 radio in a TV, the extra size may not matter, whereas in other applications, e.g., a digital camera, the space may be very important.

Because the channel select filter is very effective at removing jamming signals, the third-order intercept is not greatly affected by parts of the receiver after the filter. This allows the designer to use lower current in these circuits that could lead to some power savings. Another advantage of the superheterodyne architecture is that the channel select filter will normally remove essentially all of the adjacent and alternate channel interferers. This reduces the dynamic range requirement of the analog baseband and the ADC in the demodulator.

---

[36] Indicated in U.S. dollars.

Another advantage of the SAW filter is that the requirements for the on-chip analog baseband filters and the digital filters in the modem can be greatly relaxed. In most single-chip RFIC designs, the baseband filters are one of the largest parts of the RFIC in terms of die area. Thus, the extra cost of the SAW filter will be slightly offset by the reduced size of the baseband filters. The only baseband filtering required is to keep the noise bandwidth of the baseband below a reasonable value and to enhance the stability of the variable gain amplifiers (VGAs). Thus, the actual corner frequency of the analog baseband filters does not affect the chip performance and so would tend to help increase the yield of the chip and therefore reduce its cost. To keep high yields, other architectures like direct conversion, use calibration circuits to accurately set the corner frequency of the baseband filters using the crystal oscillator as the reference. In a superheterodyne architecture, the area for this circuit would not be needed.

If, on the other hand, the implementer desires better jamming performance, the superheterodyne architecture also allows better filtering because it is done at different frequencies. Even at low frequencies on the chip, in the 10's of MHz, the isolation between structures is finite and so the ultimate attenuation of the filter is finite due to the limited isolation in the package and in the die itself. One of the traditional reasons to use a superheterodyne receiver is to provide very high attenuation of close-in jammers, sometimes as much as 80 dB.

One of the disadvantages of the superheterodyne receiver architecture is that it often requires an image filter between the LNA and the first downconverter. If the LNA gain at the image frequency is approximately the same as it is in-band, then the noise-figure of the mixer will be reduced by 3 dB due to the additional noise at the image frequency that is mixed down to the IF. Normally, this image filter is a ceramic filter that provides 20 to 30 dB attenuation of the image frequency. However, typically, the LNA will have less gain at the image frequency than it does in-band and so the image will not noticeably affect the mixer noise figure. IEEE Std 802.15.3 does not have any requirements for rejecting a jammer outside of the allocated 2.4 GHz band. The requirements for this performance criteria will need to be determined by the implementer with input from the customer. If the requirements are relaxed, then it will be possible to achieve the image filtering requirements through the design of the LNA's bias and output matching networks rather than using a ceramic image filter.

Direct conversion

Although the direct conversion or zero IF (ZIF) architecture has received a great deal of attention in recent years as it is applied to new low-cost radios, it is really one of the oldest radio architectures. In older papers, this architecture is often referred to as a homodyne system. The appeal of the ZIF architecture is that it requires a minimum of external components, as shown in Figure 7–17. There is no image frequency for the ZIF architecture because the LO frequency is the same as the RF frequency. Because of this, there is no requirement for an image filter at the RF. The channel selectivity for the receiver is accomplished on the chip with analog baseband low-pass filters. The matched filter in a digital demodulator will also assist in filtering out the adjacent channel signals due to its low-pass nature. The I/Q down-conversion process only needs to preserve the signal quality and is not required to remove the image frequency. Thus, the requirements on I and Q phase and amplitude balance are usually more relaxed than are those for a receiver that implements image rejection with the mixers, e.g., the low IF architecture.

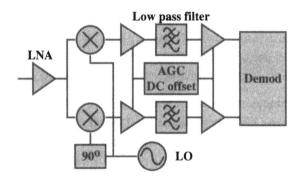

Figure 7–17: Direct conversion radio architecture

However, the ZIF architecture creates new challenges for the radio architect that are quite different from those of the superheterodyne architecture. For example, although the two-tone, third-order intermodulation test is very important for superheterodyne architectures, a single-tone, second-order intermodulation creates problems for the ZIF architecture. In the superheterodyne architecture, two tones (or one tone and an interfering signal) separated from the desired center frequency by δf and $2*\delta f$ will create a third-order product at the output of the of the amplifier at a frequency equal to the RF or at the output of the mixer at a frequency equal to the IF. The amplitude of this signal depends on the amplitude of the input signals and the cascaded third-order input intercept point (IIP3) of the receiver up to the channel select filter.

Although IIP3 performance is also an issue in the ZIF architecture, the IIP2 performance is generally a more important problem. In the ZIF receiver, signals created by mismatches in the I and Q mixers and by energy from the LO that leaks out through the RF chain and is

 The second-order products of the LNA will also create a DC voltage at the output of the LNA, but because essentially all LNA designs are AC coupled (i.e., they have a series capacitor on the output), the DC voltage is completely blocked. Thus, the IIP2 performance of a typical LNA is essentially infinite.

reflected back into the input of the mixer and create a DC voltage at the output of the mixer. Although this may be a very small offset in voltage, perhaps only 1 mV, the baseband amplifiers often have greater than 60 dB of power gain or a factor of 1000 in voltage. Thus, a 1 mV offset at the output of the mixer will turn into a 1 V offset at the output of the baseband filter/VGA combination, which would destroy the dynamic range of the ADC. The IIP2 performance of a mixer generally depends on the matching of the components in the mixer, which can be very good if the designer pays close attention to the layout.

In addition to requiring a high IP2 mixer, the ZIF architecture can deal with the DC offset problem in one of two ways. The first technique is to require high-pass characteristics from the baseband circuitry, usually by including a series capacitor. However, if the high-pass corner is too high, then the receiver performance will be degraded due to the missing signal energy that was filtered by the high-pass corner. If the high-pass corner is too low, then the filter response time will be too slow and none of the frames will be successfully received. The high-pass corner normally occurs in the 10's of kHz, which requires at least two large value capacitors, one for I and one for Q[37]. Depending on the fab process that is used for the RFIC design, these capacitors may need to be off-chip. The second technique for handling the DC offset problem is to use a DC compensation loop in the baseband filters and VGAs. The design of these loops is well-known, and the

---

[37] For a differential baseband implementation, which is typical on single chip radios, it would require twice as many components or four capacitors plus at least four pins in the package to access them.

implementation takes up a relatively small portion of the overall analog baseband die area.

However, removing the DC offset will also desensitize the receiver to signal energy around DC as well. For signals without DC content, such as orthogonal FSK or the Barker code in 802.11b, there is essentially no information at DC and this energy can be removed without affecting the performance of the demodulator. In the case of 802.15.1, the modulation is nonorthogonal FSK, and so the down-converted signal contains significant DC components. This modulation choice combined with a very short preamble made it difficult for implementers to use ZIF as an architecture for 802.15.1. Although some implementers were successful with the ZIF architecture, many opted for other methods such as low IF to work around this issue.

The direct conversion architecture also requires that the implementer create good quadrature at the RF frequency. For other standards, e.g., 802.15.1 or 802.11b, the phase and amplitude balance requirements are not very strict. The higher-order modulations of the 2.4 GHz PHY of 802.15.3, e.g., 32 and 64 QAM, on the other hand, have a relatively challenging quadrature phase and amplitude requirements to provide good signal quality on transmit (i.e., low EVM) as well as enabling low BERs for the receiver. In general, creating the quadrature signals from the VCO becomes more difficult as the frequency increases, and so trying to create this at 2.4 GHz is more difficult than creating it at a lower IF.

Walking IF

The walking IF architecture is a dual-conversion architecture that uses only one synthesizer and does not have a fixed IF frequency. Instead, because both LOs change when the channel changes, the IF will change as well. Although the IF changes with the RF frequency, the total IF bandwidth is not as large as the total RF bandwidth. The walking IF architecture, illustrated in Figure 7–18, has the advantage of using only one VCO and synthesizer while gaining many of the advantages of a superheterodyne system.

One of the key design decisions in a walking IF architecture is the choice of the divider ratio used to generate the second LO. If a divide by two is selected, then the first LO is two-thirds of the RF frequency and the second LO is one-third of the RF. This would be referred to as a two-thirds walking IF architecture. As with

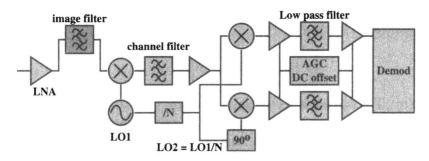

Figure 7–18: Walking IF radio architecture

a superheterodyne architecture, the designer needs to consider the frequencies where there may be interference when selecting the IF frequency as well as considering the location of possible spurs. A four-fifths walking IF would use a first LO of 1.9296 to 1.9696 GHz and have an IF center frequency that varied from 482.4 to 492.4 MHz. With the 15 MHz signal bandwidth, the image filter would need to span approximately 25 Mhz at a center frequency of 487 MHz. An advantage of the four-fifths walking IF architecture is that the divide by four circuit for the second LO would provide nearly perfect quadrature signals for the final down-conversion to the I and Q baseband signals.

The frequencies required for a walking IF architecture using low-side injection for the mixer can be determined from the following equations:

$$F_{LO1} = F_{RF}\frac{(N-1)}{N} \tag{14}$$

$$F_{LO2} = F_{IF} = \frac{F_{RF}}{N} \tag{15}$$

$$BW_{IF} = \frac{BW_{RF}}{N} \tag{16}$$

A walking IF radio can also be implemented using high-side injection for the first down-mixer. The IF frequencies and the IF bandwidth are the same in this case, only the LO is different. The LO frequency required for the walking IF using high-side injection for the mixer is:

$$F_{LO1} = F_{RF}\frac{(N+1)}{N} \tag{17}$$

Because the IF is moving in frequency, it is not possible to completely filter the adjacent channel jammers with the channel select filter. Instead, this filter only attenuates jammers in the alternate channel and those at larger frequency separation. If the second mixer has sufficient dynamic range, it is possible to use only a simple filter at the IF and perform the jammer rejection with the analog baseband and digital filters in the modem. For 802.15.3, the IF bandwidth required for the walking IF architecture, 25 MHz, is not much larger than the bandwidth required for a superheterodyne architecture, 15 MHz. On the other hand, for 802.15.1, the bandwidth for the walking IF, 17 MHz, is much larger than for the superheterodyne channel select filter, 1 MHz. So, from the point of view of the complexity of the channel select filters, the walking IF architecture is a better choice for 802.15.3 than it would be for 802.15.1.

Although the walking IF architecture does eliminate one VCO and one synthesizer as compared with a superheterodyne architecture, it has some drawbacks that are similar to the superheterodyne architecture. For example, it requires some form of image rejection, either in the form of an image filter or with an image reject mixer. The walking IF architecture also requires an IF channel select filter that is either off-chip or requires relatively large on-chip components. Because the walking IF uses two LOs, it will generate more spurious responses, and so the implementer will need to pay careful attention to the frequency plan.

Another disadvantage of the walking IF architecture is that it requires a PLL step size that is smaller than the channel spacing. For example, in 802.11b and 802.15.3, the PLL is required to support frequency step sizes of 1 MHz in order to be able to tune to the channel frequencies of 2.412, 2.437, and 2.462 GHz. On the other hand, a four-fifths walking IF PLL would have to support a frequency step size of 800 kHz in order to create the LO frequencies of 1.9296, 1.9496, and 1.9696 GHz. If the PLL uses integer division, the comparison frequency with the crystal could then be no higher than 800 kHz. This lower-comparison frequency will result in a slower lock time for the synthesizer, which can be important in frequency-hopping systems like 802.11 FHSS and 802.15.1. In the case of 802.15.3, this is not an issue as the frequency switch time is relatively long (500 µs) because the WPAN changes frequencies fairly infrequently. The smaller PLL step size can also be alleviated by using a fractional-N synthesizer instead of an integer-N synthesizer.

Low IF

An alternative to either ZIF or the relatively high IF of either the dual-conversion or walking IF architectures is to use a low IF. If the IF is equal to one-half of the RF bandwidth, then the architecture is referred to as a very low IF (VLIF) architecture. For the 2.4 GHz PHY, the IF for a VLIF architecture would be 7.5 MHz. In terms of the components required, the VLIF, as illustrated in Figure 7–19, is similar to the ZIF architecture in Figure 7–17. The key difference between the two is that in the case of the low IF or VLIF, the adjacent or alternate channel jammers will be at the image frequency and the analog baseband filters will not be able to filter out this interference. Thus, the LIF or VLIF architecture needs to provide a much higher level of image rejection.

In Figure 7–19, the image rejection is obtained using complex filters. A filter that has a real time-domain response will have an even amplitude response in frequency; i.e., negative frequencies are attenuated the same amount as the positive frequencies. A complex filter, on the other hand, can be constructed to have an odd response in the frequency domain; i.e., the positive frequencies will be passed, whereas the negative frequencies are attenuated or vice versa. The complex filter is realized by feeding part of the I channel back in to the Q channel and vice versa. The filters are more difficult to implement than is the standard analog low-pass filter due to the higher complexity, and most designers are not familiar with the techniques required to implement them. However, some companies have implemented this approach in 802.15.1 receivers.

As an alternative to complex filters, the receiver can also reject the image frequencies by performing a complex multiplication on the I and Q channels, as shown in Figure 7–20. In this implementation, the complex IF signal is multiplied by a complex LO that converts the IF signal to baseband. This multiplication shifts the baseband spectrum from the IF to DC without folding the image back into the pass band. Normally, this multiplication is performed digitally in the modem, e.g., with a CORDIC[38] algorithm, after the baseband I and Q channels have been sampled with an ADC. Because the final down-conversion is performed in the digital domain, the yields from this part of the system are very high due to the repeatability of modern digital IC processes.

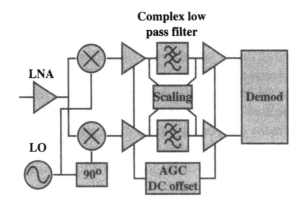

Figure 7–19: Low IF radio architecture

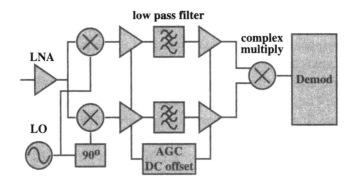

Figure 7–20: Alternate low IF radio architecture

---

[38] CORDIC is an acronym for COordinate Rotation DIgital Compute, which is a class of shift-add algorithms for rotating vectors in a plane.

The complex multiplication requires a total of four real multiplications and two additions which can be seen by multiplying two complex numbers, e.g:,

$$\gamma_1 \times \gamma_2 = (\alpha_1 + j\beta_1) \times (\alpha_2 + j\beta_2) = \alpha_1\alpha_2 - \beta_1\beta_2 + j(\alpha_1\beta_2 - \alpha_2\beta_1) \quad (18)$$

Because of this property, the complex multiplication is sometimes represented by four digital mixers and could be implemented with four analog mixers as well.

Another method to obtain the image rejection required for a low-IF receiver is to use the Weaver architecture [B30], as illustrated in Figure 7–21. In this architecture, two sets of analog mixers are used to down-convert the signal to a low IF without the image. The demodulator will need to either subsample the IF or mix it down to baseband. As the image has been removed, the demodulator does not need to implement a full complex multiplication but instead only needs to perform an I/Q down-conversion. One of the advantages of this architecture is that only one ADC is required to digitize the signal for the demodulator.

The low-IF receiver is more challenging to design than is the ZIF receiver due to the increased image rejection requirements. However, the low IF does not suffer from the DC offset problems that complicate the ZIF design. In addition, it does not require external filters, as in the case of the superheterodyne and walking IF architectures.

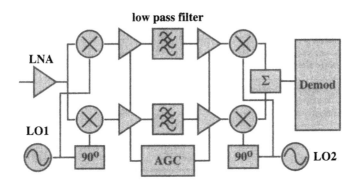

Figure 7–21: Low IF using the Weaver architecture

Summary of radio architectures

The choice of a radio architecture depends on many different factors. Part of the considerations are the characteristics of the modulation type and frequency. Perhaps equally important, however, are the skills of the RFIC design team that is going to do the actual design. No radio architecture will simultaneously solve all

> It is the author's opinion that selecting a particular radio architecture will not remove the pain in a design; it only allows you to determine where you will get it. TANSTAAFL is especially true in RF design. (TANSTAAFL = There ain't no such thing as a free lunch.)

of the problems; instead, each one has a different area in which the designers will find difficulty.

Table 7–24 summarizes some of the characteristics of each of the radio architectures discussed in this section.

Any one of the architectures discussed in this section can be used to design a low-cost 802.15.3 standard 2.4 GHz PHY, although the architectures that require a minimum of external filters, i.e., the ZIF and LIF/VLIF, will probably be the ones most likely to be implemented for the low-cost, relaxed performance required for 802.15.3.

Table 7–24: Comparison of some receiver architectures

Characteristic	Super-het	Walking IF	ZIF	LIF/VLIF
Off-chip filters?	Yes[a]	Yes[a]	No	No
Off-chip image filter?	Yes[b]	Yes[b]	No	No
Strong image reject performance?	No	No	No	Yes
DC offset correction required?	No	No	Yes	No
Complexity	Highest	High	Lowest	Low
Jamming resistance	Best	Good	Fair	Fair

[a] In the case of the superheterodyne and walking IF architectures, the channel select filter can be replaced with on-chip filtering and baseband low-pass filters. However, this approach requires substantially more dynamic range out of the second set of mixers and discards some of the performance advantages of these architectures.
[b] Depending on the actual design and performance requirements, these filters could be implemented on the chip.

Chapter 8 Interfacing to 802.15.3

IEEE Std 802.15.3 specifies MAC and PHY layers to provide interoperability between compliant DEVs that implement the standard. In addition to the requirements for the MAC and PHY, a typical implementation also includes, at least conceptually, a variety of management entities that control the MAC and PHY and provide interfaces to allow higher layers to control the operation of the MAC and PHY. Each of these management entities provides a layer management service interface, referred to as a service access point (SAP) through which the management entity is controlled. Figure 8–1 provides an illustration of the various management entities and the SAPs that are used for communication and control.

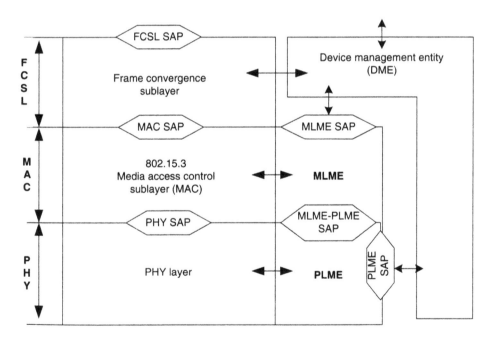

Figure 8–1: Reference model used in IEEE Std 802.15.3

The reference model in Figure 8–1 represents only one possible arrangement of the layers and management entities. An implementer is only required to provide a minimum interface at the frame convergence sublayer (FSCL) SAP and at the PHY antenna. None of the other interfaces are mandatory, and so they can be implemented in a manner that is the most efficient for the implementer. This model is relevant to the text and descriptions used in the standard as it provides a reference for the terms and divisions that were used while developing the standard.

The device management entity (DME) is intended to provide "glue" between the various layers and to implement all of the undefined control functions for the various layers. The exact function of the DME is not specified in the standard. Instead, the DME provides a location where all of the undefined procedures of the standard exist. In general, if the actions by the DEV or the PNC are fully-defined in the standard, then they will exist in either the PHY or the MAC. Any implementation-dependent functionality in the standard resides, at least conceptually, in the DME. The only exception to this rule is the channel time allocation (CTA) algorithm in the PNC. Instead of passing the request for channel time to the DME, the standard indicates that PNC's MAC/MLME makes the decision on allocating channel time, but it does not indicate the method that will be used to make the decision. The Task Group wanted to keep the CTA algorithm as an implementation dependent function for a couple of reasons. First, leaving this outside of the standard allows companies to innovate in this area to provide differentiation to products that would otherwise be functionally the same. A PNC with a good CTA algorithm will be able to better maintain the QoS in piconet and to provide greater overall throughput. Both of these are qualities of the piconet that are visible to the end user in the form of the performance of their applications.

The second reason that the CTA algorithm is not specified in the standard is that no member of the group proposed one for inclusion in the standard. Most companies consider their CTA algorithm to be a proprietary advantage in developing solutions, and so they do not want to publish their methods. In addition, even if CTA algorithms had been proposed, it is unlikely that the task group would have been able to come to agreement on one algorithm to require in the standard. The "correct" algorithm depends on the type of traffic that the PNC expects to support, the level of complexity that can be supported by the price of the implementation, and so on.

Although some interfaces between the management entities are defined in the standard, other key interfaces are only implied and are not specified in the standard. For example, the FSCL to DME interface is not defined, and yet this interface is required for the creation of streams. Isochronous data come into the FCSL, which maps them into stream indices, but it is the DME in the standard that requests, modifies, and terminates all isochronous stream allocations.

Another missing interface is the interface between either the DME or the FCSL and the higher layers to enable them to request an isochronous stream connection. The current 802.2 FCSL only defines an interface to pass individual frames that do not have a relationship to any of the other frames that might be passed to the FCSL. In the case of a stream, the frames that enter the FCSL really belong to one of the allocated streams in the superframe, but there is not an interface that allows the FCSL to be able to map the incoming data frames into the existing streams in the piconet.

The standard also does not specify the interface between the MLME and the MAC or between the PLME and the PHY. In modern implementations, the layer management entities are tightly bound to their respective layers, and there is no need to specify a separate interface that will never be exposed in practical implementations.

The SAPs that are defined in the standard are the following:

• FCSL SAP
• MAC SAP
• PHY SAP
• MLME SAP
• PLME SAP
• MLME-PLME SAP

The PLME SAP and the MLME-PLME SAP use identical primitives, and so they are viewed instead as a single SAP, referred to as the PLME SAP, that can be used either by the MLME or the DME to control the behavior of the PHY. Of course, the actual implementation does not have to implement separate interfaces or to even support the PLME SAP as an external interface.

The MLME SAP in the standard is a relatively detailed interface (it has 22 subclauses, most with more than one subclause describing a different primitive). In fact, it is much more detailed than is the equivalent interface defined for 802.11. The Task Group spent an enormous amount of time on specifying this interface, which is optional anyway, to enable the formal verification of the standard using methods like the structured definition language (SDL). SDL simulators allow a protocol to be modeled and tested to find deadlocks, live locks, and other problems in the protocol in a formal manner. SDL modeling tools also allow the generation of message sequence charts (MSCs) from the SDL model as well as generating object code that can be used to instantiate the protocol. The MLME SAP can be used with an SDL modeling tool to model and test the 802.15.3 protocol as it is written in the standard (with the exception of the CTA algorithm, which is in the MLME but is not specified).

Quite a bit of effort was spent in trying to keep the MLME SAP up to date with the protocol description contained in the clauses for the frame formats (Clause 7) and the MAC functional description (Clause 8) in the standard. The MLME SAP can be viewed as being derived from the other two clauses rather than introducing its own functionality. Because it is a derived clause, many of the technical comments received in the letter ballots had to do with changes in the protocol that were specified in Clause 7 and Clause 8 but were not reflected by changes in Cause 6. In order to create an interoperable, 802.15.3-compliant DEV, the implementer only needs to refer to Clause 7, Clause 8, Clause 9 (if security is implemented), and Clause 11. The information in Clause 6 for the MLME SAP is only a guide.

The standard explicitly states that the interfaces described in the layer management clause (Clause 6) do not have to be implemented as detailed in the standard if they are not exposed as external interfaces. Even if the interfaces are exposed, the implementer "should" support the primitives defined in the clause, but they are not required to support those primitives. If the implementer does support the primitives defined in the layer management clause, they are not restricted from creating additional primitives that may be required to complete the implementation or to add vendor specific functionality to the MAC.

THE PIBs AND THEIR INTERFACE

The PAN information base (PIB) contains information that is similar to that found in the management information base (MIB) used in other standards like 802.11. The standard uses a different name to differentiate between the data that might be managed over a network using SNMP, i.e., the MIB, versus information that is controlled only by the local DEV and its applications, i.e., the PIB. The PIB is not intended to be used for centralized management because the assumption of IEEE Std 802.15.3 is that the DEVs often operate in piconets that do not have connections with the larger networks, e.g., LANs, MANs, or WANs.

The information in the PIB can be written using either the MLME-SET.request primitive or the PLME-SET.request primitive. Likewise, the MLME-GET.request primitive or the PLME-GET.request primitive are used to retrieve information from the PIB. These primitives use the same syntax, and so they are generically referred to in the standard as XX-GET and XX-SET. The MLME primitives retrieve or set information in the MAC PIB, whereas the PLME versions of the primitive are used to retrieve or set information in the PHY PIB. The MAC and PHY PIBs can be thought of as being contained in MLME and PLME, respectively. Thus, the PIBs contain information that will typically be used by either the MAC or the PHY in its normal operation.

The PIB entries may be either read only, write only, or read-write, depending on how they are defined in the PIB tables in the standard. If the XX-SET.request primitive attempts to write a read-only PIB entry, it will generate an error. Likewise, if the XX-GET.request primitive tries to retrieve the value of a write-only PIB entry, it will also generate an error. A PIB entry that has the characteristic of write-only restricts the operation of the XX-GET primitive. The PLME or MLME can read or write this PIB entry; it is only the DME that is forbidden to read it.[39]

---

[39] Some engineers at a semiconductor company were almost successful in releasing an official data sheet for "write-only memory" as a joke, but were caught by a unusually vigilant marketing staff.

MLME SAP

The MLME SAP, as described earlier, provides a relatively complete list of primitives that can be used to model the defined behavior of the standard. Because these optional primitives are linked to the normative behavior of the frame exchanges in the wireless medium, they are also referenced in the MAC functional description (Clause 8 in the standard) in both the text and the MSCs. However, the presence of these primitives in the text of Clause 8 or in the MSCs of Clause 8 should not be viewed as requiring a particular implementation of this interface. As it is stated in Clause 6, this interface is optional and implementers are free to create any division of functionality within the MAC and to create an interface that is appropriate to that division.

Therefore, the MLME SAP should not be viewed as an indication of the correct architecture for the MAC or as the point at which the MAC is split into hardware and software. An implementer may create a product in which essentially all functions of the MAC and DME are in hardware, whereas another implementer may realize essentially all of the functions in software. Either implementation is correct as long as it adheres to the frame formats in Clause 7 and the sequence of frames sent in the wireless medium as defined in Clause 8 of the standard. The primitives provided by the MLME SAP are summarized in Table 8–1. Because the MLME SAP is simply a derived interface, the functionality provided by these has already been described in Chapter 4, Chapter 5, and Chapter 6 of this handbook.

The description of the MLME SAP primitives in Clause 6 in the standard makes the mistake of trying to also describe some of the functionality of the DME, which is outside of the scope of the standard. All of the MLME SAP primitives in Clause 6 include subclauses for "when generated" and "effect of receipt." However, these subclauses are only relevant when they refer to the operation of the MAC and cannot apply to the operation of the DME, which is not supposed to be specified. Accordingly, the XX.request and XX-response primitives should have only the "effect of receipt" subclause because this describes what the MAC will do when it receives these primitives. In a similar fashion, the XX.indication and XX.confirm primitives should only specify the "when generated" subclause and not the "effect of receipt" subclause because these indicate an action that the MAC has taken. The DME's response to the XX.indication and XX.confirm primitives is not

Table 8–1: Summary of MLME primitives

Name	Request	Indication	Response	Confirm
MLME-RESET	X	–	–	–
MLME-SCAN	X	–	–	X
MLME-START	X	–	–	X
MLME-START-DEPENDENT	X	–	–	X
MLME-SYNCH	X	–	–	X
MLME-ATP-EXPIRED	–	X	–	–
MLME-ASSOCIATE	X	X	X	X
MLME-DEV-ASSOCIATION-INFO	–	X	–	–
MLME-DISASSOCIATE	X	X	–	X
MLME-REQUEST-KEY	X	X	X	X
MLME-DISTRIBUTE-KEY	X	X	X	X
MLME-DE-AUTHENTICATE	X	X	–	X
MLME-MEMBERSHIP-UPDATE	X	–	–	–
MLME-SECURITY-ERROR	–	X	–	–
MLME-SECURITY-MESSAGE	X	X	–	X
MLME-PNC-HANDOVER	X	X	X	X
MLME-NEW-PNC	–	X	–	–
MLME-PNC-INFO	X	X	X	X
MLME-SECURITY-INFO	X	X	X	X
MLME-CREATE-ASIE	X	–	–	X
MLME-RECEIVE-ASIE	–	X	–	–
MLME-PROBE	X	X	X	X
MLME-ANNOUNCE	X	X	–	X
MLME-PICONET-SERVICES	–	X	X	X
MLME-CREATE-STREAM	X	–	–	X
MLME-MODIFY-STREAM	X	–	–	X
MLME-TERMINATE-STREAM	X	X	–	X
MLME-MULTICAST-RX-SETUP	X	–	–	–
MLME-CHANNEL-STATUS	X	X	X	X
MLME-REMOTE-SCAN	X	X	X	X
MLME-PICONET-PARM-CHANGE	X	–	–	X
MLME-TX-POWER-CHANGE	X	X	–	X
MLME-PS-SET-INFORMATION	X	–	–	X
MLME-SPS-CONFIGURE	X	–	–	X
MLME-PM-MODE-CHANGE	X	–	–	X
MLME-PM-MODE-ACTIVE	–	X	–	–

defined in the standard. When this mistake was realized, it was too late in the process to clean it up without potentially introducing more technical errors in the standard. Instead, this clean up of the layer management clause will have to wait until an amendment project authorization request (PAR[40]) is approved for IEEE Std 802.15.3.

With those caveats in mind, the MLME SAP does provide some additional insight into the functions that are fully defined in the standard versus functions that are either implementation specific or are outside of the scope of the standard. For example, consider the primitives associated with power management, MLME-PM-XX, MLME-SPS-XX, and MLME-PS-XX. The standard includes the descriptions of the primitives that are generated by the power save DEV, the XX.request and XX.confirm, because the DEV determines its power-save mode characteristics in a manner that is outside of the scope of the standard. The DEV will want to consider the characteristics of its PHY in terms of AWAKE state versus SLEEP state power drain as well as the potential traffic that it could be sending or receiving.

On the other hand, the PNC's participation in the power-save mode process does not have any MLME SAP primitives associated with it. A quick examination of the MSCs in the power management subclause of Clause 8 will show that there is absolutely no interaction with the PNC DME. In fact, the PNC DME appears in only one of the MSCs in the subclause, and in that instance, it could have been deleted as well. The behavior of the PNC for supporting the power-save modes is very straightforward and is well defined in the standard. Most of its activity is spent setting bits in IEs, placing these in the beacon, and updating the next wake beacon from time to time. The PNC will generally accept all requests by the DEVs (unless they make a mistake or get mixed up in the order that the frames are supposed to be sent in the wireless medium). Thus, no MLME SAP primitives are required by the PNC to support the power-save mode DEVs.

---

[40] A PAR is a project authorization request, the means by which standards projects are started in the IEEE Standards Association. PARs define the scope, purpose, and contact points for the new project.

PLME SAP

The PLME SAP represents another conceptual interface in the standard. This interface provides higher-level capabilities to control the PHY. A total of five primitives in three classes are defined for this SAP. The list of supported primitives is given in Table 8–2.

Table 8–2: Summary of PLME SAP primitives

Name	Request	Indication	Response	Confirm
PLME-RESET	X	–	–	X
PLME-TESTMODE	X	–	–	X
PLME-TESTOUTPUT	X	–	–	–

The PLME primitives are pretty self-explanatory. The PLME-RESET causes the PHY to clear any settings and set the PHY in the off state (i.e., neither transmitting nor receiving). The other two primitives are provided to enable testing of the PHY without enabling the MAC functionality. The PLME-TESTMODE primitive is used to set up the PHY for transmitting one of three test patterns, and the PLME-TESTOUTPUT primitive instructs the PHY to begin the transmission. These primitives only provide a small portion of the test characteristics that would normally be implemented in a design. An actual implementation would also have receiver test modes and possibly a set of loop-back mode for self-testing of the PHY.

MAC SAP

The MAC SAP provides an interface between one or more FCSL layers and the MAC sublayer. The FCSL takes frames passed from higher layers and identified with source and destination addresses and maps this data into the SrcID, DestID, and stream index to be used in transmitting the data. The MAC SAP then provides six primitives, three each for isochronous and asynchronous data, which are summarized in Table 8–3.

Table 8–3: Summary of MAC SAP primitives

Name	Request	Indication	Response	Confirm
MAC-ASYNC-DATA	X	X	–	X
MAC-ISOCH-DATA	X	X	–	X

The asynchronous primitives specify the target and destination IDs of the data, whereas the isochronous primitives use only the stream index. The stream index is sufficient for isochronous streams because each assigned stream index applies to only one SrcID/DestID pair. Asynchronous data, on the other hand, shares a single stream index, and so it is not possible to determine the SrcID/DestID pair from the stream index alone. Each of the XX.request primitives also includes a TransmissionTimeout parameter that limits the length of time that a DEV will attempt to send the frame before it gives up on it and discards the frame.

PHY SAP

In IEEE Std 802.11, the PHY is split into two sublayers, the physical layer convergence protocol (PLCP) and the physical medium-dependent (PMD) sublayers. The initial IEEE Std 802.11 supported three different, non-interoperable PHYs,[41] and so the working group used the PLCP to provide an interface between the three incompatible layers and the single MAC.

Instead of this approach, IEEE Std 802.15.3 defined only the PHY layer and did not explicitly split it up into separate sublayers. The functionality equivalent to the PLCP and PMD sublayers of 802.11 is included in the 802.15.3 PHY layer; it just is not split into two different sublayers. However, this does not mean that IEEE Std 802.15.3 does not support multiple PHYs. The PHY SAP interface provides sufficient abstraction so that other future PHYs can be implemented with the 802.15.3 MAC. This process is already underway with the development of 802.15.3a, a higher-rate PHY for the 802.15.3 MAC.

---

[41] IEEE Std 802.11 now defines a total of nine different PHYs, none of which are mandatory and six of which, 802.11 DSSS, 802.11b (2 PHYs), and 802.11g (3 PHYs) are interoperable at some level.

The development process of this new PHY uses the PHY SAP to define the demarcation between the MAC and the PHY. For example, all of the proposals will be evaluated based on the throughput that is delivered at the PHY SAP interface. This includes the overhead for the PHY frame, on-channel timing, and any forward error correction (FEC) if it is included in the proposals.

The PHY SAP provides primitives in nine categories that relate to the operation of the PHY. The primitives that are provided by the PHY SAP are summarized in Table 8–4.

Table 8–4: Summary of PHY SAP primitives

Name	Request	Indication	Response	Confirm
PHY-DATA	X	X	–	X
PHY-TX-START	X	–	–	X
PHY-TX-END	X	–	–	X
PHY-CCA-START	X	–	–	X
PHY-CCA-END	X	–	–	X
PHY-CCA	-	X	–	–
PHY-RX-START	X	X	–	X
PHY-RX-END	X	X	–	X
PHY-PS	X	–	–	X

The PHY-TX-START, PHY-TX-END, PHY-RX-START, and PHY-RX-END are self-explanatory; they provide methods to begin and end transmissions and receptions. The PHY-RX-START and PHY-RX-END primitives also provide indications because the DEV will not necessarily know when a frame that is intended for it will arrive at the receiver. In addition, the PHY-RX-END will return additional information regarding the attempt to receive the frame. For example, it will indicate if the signal was lost in the middle of a frame, that the format of the frame was wrong, or that the frame was sent with an unsupported data rate.

The data rate and MAC header are passed to the PHY with the PHY-TX-START.request and are received from the PHY with the PHY-RX-START.indication. The PHY-RX-START.indication also returns the received signal strength indication (RSSI) measured during the MAC and PHY headers, whereas the PHY-RX-END.indication returns the link quality indication (LQI).

Once the PHY is in either transmit or receive mode, the PHY-DATA primitive is used to either pass data to the PHY to be sent or to pass data from the PHY to the MAC as it is received. The data are passed in this primitive one octet at a time, but an implementation may choose to use a different method to transfer the data between the PHY and the MAC.

The CCA primitives are used in the CAP as well as in the retransmission timing. The MAC begins the CCA process by sending the PHY-CCA-START.request and ends the process by sending the PHY-CCA-END.request. While the process is running, the MAC will receive PHY-CCA.indications periodically that indicate the current status of the channel, either busy or idle.

THE FCSL

The frame convergence sublayer (FCSL[42]) was developed by the 802.15.3 Task Group when it became apparent that the IEEE 802.2 LLC would not provide an interface that could support all of the types of multimedia data that IEEE Std 802.15.3 was intended to provide. The IEEE 802.2 LLC is a data-centric layer that works well for asynchronous data traffic in a network. However, it was not designed with streams of isochronous data in mind. There are other higher layer standards, e.g., IEEE Std 1394 and USB, that do provide support for the creation, modification, and termination streams of isochronous data. Because an interface for data and legacy equipment is a requirement for the standard, IEEE Std 802.15.3 needed to provide the standard IEEE 802.2 LLC interface. However, to support alternative methods of negotiating stream connections, IEEE Std 802.15.3 created a thin layer, the FCSL, that provides a method for multiple LLC layers to interface with the 802.15.3 MAC. This interface is illustrated in Figure 8–2.

---

[42] For the benefit of the curious reader, the acronym is FCSL instead of FCS because the frame-check sequence preceded the naming of the FCSL by at least a year. The FCSL was originally named the service-specific convergence sublayer (SSCS), but it was renamed in one of the Working Group letter ballots.

The FCSL performs the following functions when it receives a protocol data unit (PDU) from an upper layer:

a) Maps the source addresses and destination address into a SrcID and Dest-ID.

b) Uses the priority and other information to determine the correct stream index.

c) If the frame needs a stream allocation, it sends a request to the DME to create a new stream.

d) Delivers the PDU to the MAC SAP either as an asynchronous MSDU with a SrcID and DestID or as an isochronous MSDU with an associated stream index.

When the FCSL receives an MSDU from the MAC SAP, it reforms the following actions:

a) Maps the received DestID and SrcID or the stream index into a source address and a destination address.

b) Delivers the PDU to the appropriate higher layer.

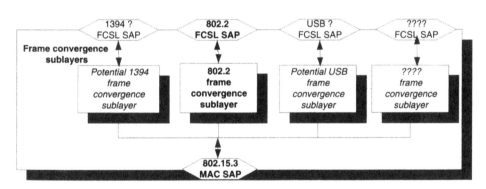

Figure 8–2: Frame convergence sublayer model

IEEE Std 802.15.3 defines only one FCSL, the IEEE 802.2 FCSL. Other FCSLs may be defined as a a part of a later MAC enhancement PAR, via an industry SIG or by another standards group, e.g., IEEE Std 1394. In fact, the IEEE 1394[43]

---

[43] The IEEE Std 1394 is often referred to as FireWire™.

Trade Association (TA) has recently started a project (see [B21]) to write a protocol adaptation layer (PAL) for 802.15.3. A 1394 PAL is roughly equivalent to the 802.15.3 FCSL. When this project is completed, it should provide a support for the upper layer parts of 1394 on the 802.15.3 MAC and PHY.

The IEEE 802.2™ [B11] FCSL provides QoS support by either using a best effort scheme or by using the hierarchical IEEE 802.1D™ [B10] QoS priority scheme. The best effort scheme treats all frames the same; i.e., there is no attempt to deliver some frames with less delay than other frames. The 802.1D QoS priority scheme defines eight possible priority levels with seven defined traffic types. The 802.1D™ traffic types are listed in Table 8–5.

Table 8–5: IEEE Std 802.1D traffic types

User priority	Traffic type	Used for	Comments
0 (default)	Best effort (BE)	Asynchronous	Default level
1	Background (BK)	Asynchronous	–
2	–	Spare	Currently not assigned
3	Excellent effort (EE)	Isochronous	For valued customers
4	Controlled load (CL)	Isochronous	Traffic will have to conform to some higher protocol layer admission control
5	Video (VI)	Isochronous	< 100 ms delay and jitter
6	Voice (VO)	Isochronous	< 10 ms delay and jitter
7	Network control (NC)	–	–

The IEEE 802.2 FCSL provides only three primitives to send and receive data, MA-UNITDATA.request, MA-UNITDATA.indication, and MA-UNITDATA-STATUS.indication. The MA-UNITDATA.request primitive is used to send a PDU, whereas the MA-UNITDATA.indication is used to notify the higher layers that a PDU has been received. The MA-UNITDATA-STATUS.indication is used to report the status of the transmission attempt for the PDU from the immediately-preceding MAC-UNITDATA.request.

Chapter 9 Coexistence mechanisms

INTRODUCTION

The wireless medium is, by its nature, a shared medium. Because of this, the development of the 802.15.3 standard considered various methods of coexistence and adopted various methods and recommendations that enable compliant devices (DEVs) to achieve a level of coexistence with other IEEE standards. There are many definitions that have been proposed for coexistence, but for the purpose of IEEE Std 802.15.3, coexistence is defined as "The ability of one system to perform a task in a given shared environment where other systems have an ability to perform their tasks and may or may not be using the same set of rules." This definition recognizes that the impact of two systems on each other is not a single number, but rather a group of performance curves that depend on the physical configuration of the networks, the amount and type of traffic on the networks, and the specific choices made by an implementer.

In general, as soon as any two wireless systems that are operating in the same general frequency are brought into proximity with each other, there is some degradation in the throughput of at least one of the systems. Thus, it is not realistic to expect that there would be no effect on either system by the presence of another. Even if the systems cooperate with each other, the methods used only minimize the reduction in throughput, they do not to eliminate it entirely. For example, if the networks share the medium agreeing to a time-sharing protocol, e.g., time division multiplexing (TDMA), the throughput of the systems is reduced by the amount of time that is allocated to the other network. However, this sharing can improve the aggregate throughput as compared with the case when the two networks have not synchronized their timing.

COEXISTENCE TECHNIQUES IN 802.15.3

The coexistence approach for 802.15.3 focused on a variety of areas, ranging from the selection of the PHY channels to the ability to coordinate timing with networks that are not 802.15.3 compliant. In all, 802.15.3 identified a total of eight different techniques that can be used to improve the level of coexistence among 802.15.3 networks and with other types of wireless networks. These techniques include the following:

- Passive scanning
- Dynamic channel selection
- The ability to request channel quality information
- Link quality and RSSI
- A channel plan that minimizes channel overlap
- Transmit power control
- Lower impact transmit spectral mask
- Neighbor piconet capability

Although all of these techniques are in 802.15.3, not all of them are applicable to improve the level of coexistence with other wireless networks. Table 9–1 lists the coexistence techniques and indicates if they are applicable to a particular IEEE standard. In the table and throughout this chapter, 802.11b also implies 802.11 DSSS and 802.11g because of the similarities in the PHYs. Likewise, 802.15.1 also includes the Bluetooth specification because it is the IEEE standardization of that specification and includes 802.11 FHSS by virtue of the similarities of the PHYs.

Table 9–1: Applicability of coexistence techniques to IEEE standards

Technique	802.11b	802.15.1
Passive scanning	X	X
Dynamic channel selection	X	X
Requesting channel quality information	X	X
Link quality and RSSI	X	X
Channel plan to minimize overlap	X	–
Transmit power control	X	X
Lower impact transmit spectral mask	X	X
Neighbor piconet capability	X	–

Passive scanning

When an 802.15.3 DEV wants to either create or join a piconet, it is required to passively scan a potential channel before attempting to start a piconet. Passive scanning means that the process of starting or joining a piconet will not cause an interruption to existing networks. If the DEV is trying to start a new piconet, it will be looking for a channel that is relatively quiet. This increases the probability that it will not start up a piconet on a channel that is already in use by another wireless network. In addition, an 802.15.3 DEV may be able to determine that an 802.11b WLAN is present in the area. In this case, the 802.15.3 DEV will avoid using the channel in use by the 802.11b WLAN.

The ability to request channel quality information

In order to improve coexistence, it is useful to have information about the amount of interference in each of the channels as well as information about other wireless networks present in the channel. In the case of 802.15.3, not only does the PNC sense the channel in its area, but it is also capable of asking any other DEV to respond with its own estimate of the channel status. This command is useful for detecting coexistence problems in remote DEVs by the PNC or other DEVs that are unable to detect an interference environment (for example, during a passive scan).

Dynamic channel selection

The 802.15.3 piconet coordinator (PNC) periodically requests channel status information from the DEVs in the piconet. The DEVs respond with statistics about how many frames have been successfully transmitted or received. This information allows the PNC to determine that the channel is having problems, as it would when an 802.11b WLAN is present. If the PNC determines that the channel is getting too much interference, it searches the other channels to find one that has a lower level of interference. If the PNC finds a channel with less interference, then the PNC moves the piconet to a quieter channel. Thus, if an 802.11b network is present, the 802.15.3 piconet would change channels to avoid interfering with 802.11b. Likewise, if an 802.15.1 PAN is present and is using adaptive frequency hopping to avoid a portion of the spectrum, the PNC could move the piconet to the quiet portion of the spectrum.

Link quality and RSSI

The 2.4 GHz PHY specifies that a DEV returns the RSSI and, for the higher order modulations, an estimate of the link quality. The RSSI provides an estimate of the strength of the received signal, which can be used for transmit power control. The RSSI combined with the link quality indication (LQI) provides a method to differentiate between low signal power and interference causing the loss of frames. For example, if the RSSI is low and frames are being lost, then the cause is low receive power. On the other hand, if the RSSI is relatively high, but the LQI is low, that would indicate the possibility of interference in the channel.

Channel plan that minimizes channel overlap

The channel plan for the 2.4 GHz PHY balances the requirement of four simultaneous piconets with the desire to coexist with other wireless standards, such as 802.11b. To do this, two channel sets are available (see "Channel plan" on page 205). The high-density channel plan was selected to allow four piconets to operate in relatively close proximity. The reduced transmit spectral mask of 802.15.3 allows one more non-overlapping channel as compared with 802.11b. The 802.11b coexistence channel plan, on the other hand, only allocates three channels so that it matches up with the frequencies for 802.11b.

The PNC chooses the channel plan that it will use for the piconet. If the PNC does not detect an 802.11b WLAN, then one of the channels in the high-density channel plan would be used for its piconet. On the other hand, if the PNC detects an 802.11b WLAN, then it would choose the 802.11b coexistence channel plan. The reason for this is that each of the two center channels in the high-density channel plan would overlap two 802.11b channels. The 802.11b coexistence channel plan aligns the channels so that 802.15.3 operation in one of these channels would only affect a single 802.11b channel. Table 9–2 shows how the 802.11b coexistence channel plan for 802.15.3 compares with the channels for 802.11b in North America and Europe. By choosing the "quietest" one, the 802.15.3 piconet would minimize its impact on the 802.11b networks.

If the PNC has already selected the high-density channel plan and is operating on one of the two inner channels, i.e., either channel 2 or 4, it is able to switch channel plans if it detects an 802.11b WLAN in its operational area. In this case, the PNC would use the piconet parameter change process to move the piconet to a new channel in the 802.11b coexistence channel plan. Note that channel 1 or channel 5 is the same in both plans, so piconets operating on either of these two channels would not necessarily have to change channels if an 802.11b WLAN was detected (unless it is on the same channel).

Table 9–2:   Comparison of 802.15.3 and 802.11b channel center frequencies

802.15.3 channel center frequency	802.11b North American channel center frequency	802.11b European channel center frequency
2.412 GHz	2.412 GHz	2.412 GHz
2.437 GHz	2.437 GHz	2.442 GHz
2.462 GHz	2.462 GHz	2.472 GHz

Transmit power control

IEEE Std 802.15.3 provides three methods for controlling transmit power. The first method is that the PNC is able to set a maximum power level for the beacon, CAP, and directed MTSs. For the 2.4 GHz PHY, this can be set as low as 0 dBm. This allows 802.15.3 piconets to reduce the interference they create for other networks while maintaining the operation of the piconet.

The second transmit power control method allows DEVs in a CTA to request a change in the transmit power of the remote DEV for that link. The originating DEV sends a Transmit Power Change command to the target DEV that tells the target DEV the amount to change its power up or down, depending on the received signal power. Thus, two DEVs that are relatively close to each other are able to both save power and reduce the interference to other networks while maintaining a high-quality link. The third method for controlling the transmitter power is that DEVs are allowed to change their transmit power based on their own estimation of the channel.

Lower impact transmit spectral mask

The transmit spectral mask for 802.15.3 was chosen as a balance of complexity, performance, and coexistence. A wider transmit spectrum reduces the filter requirements, thereby reducing cost. It also results in a slightly better BER performance for a given SNR. However, a wider transmit spectrum also causes more problems to networks operating in adjacent and alternate channels. It also makes the receiver more susceptible to other interference sources. Because of this, this standard chose a narrower transmit spectral mask than is used for either 802.11b or 802.11g, as shown in Figure 9–1.

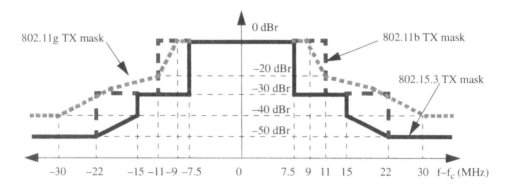

Figure 9–1: 802.15.3 and 802.11b transmit spectral mask

As the figure shows, the 802.15.3 transmit spectral mask is significantly smaller than the one required for either 802.11b or 802.11g. The smaller mask reduces the interference that 802.15.3 DEVs cause other wireless networks. It also allows narrower receive filters for 802.15.3 DEVs, which reduces the interference caused by other wireless networks in an 802.15.3 piconet.

Neighbor piconet capability

The neighbor piconet capability allows a DEV, which may not be fully 802.15.3 compliant, to request time to operate a network that is collocated in frequency with the 802.15.3 network. Although current 802.11b radios do not implement this functionality, it is relatively easy to build 802.11b radios that could support enough of a subset of 802.15.3 to request the neighbor piconet capability. One reason that this can be done is that the 802.15.3 PHY has many similarities with 802.11b that make it easier to build dual-mode radios.

One of the similarities is that 802.11b and 802.15.3 share the same frequency band, which makes interoperability of radio modules much simpler. Also, the 802.15.3 PHY layer uses 11-Mbaud DQPSK modulation for the base rate, which is the same as the chip rate and modulation for 802.11b. However, 802.11b uses either a Barker code, CCK or PBCC as a spreading code, none of which are a part of the 802.15.3 standard. The 802.15.3 PHY was also chosen with the same frequency accuracy, allowing the reuse of reference frequency source and frequency synthesizers. In 802.11b coexistence mode, 802.15.3 uses the same frequency plan as 802.11b. In addition, the synthesizers that would normally be used in either 802.11b or 802.15.3 would be capable of 1-MHz freqency step size and so would be capable of supporting either frequency plan. The RX/TX turnaround time is also the same for both protocols. However, the TX/RX turnaround for 802.11b is 5 µs vs. 10 µs for 802.15.3, which could have an impact on the architecture of a dual-mode radio. To summarize, the similarities between 802.15.3 and 802.11b are as follows:

- DQPSK modulation
- 11-Mbaud symbol (chip) rate.
- Frequency and symbol timing accuracy of +/– 25 ppm.
- RX/TX turnaround time
- Power ramp up/down
- Frequency plan

Some of the differences include:

- Barker, CCK or PBCC spreading code
- Power spectral density
- Performance criteria (e.g. sensitivity, jamming resistance, etc.)
- TX/RX turnaround time.
- PHY preamble, header, frame structure
- MAC

A dual-mode 802.11b/802.15.3 access point could request a CTA for a neighbor piconet to share time with an existing 802.15.3 piconet. The AP would use the

PCF and NAV to set aside time for the operation of the 802.15.3 piconet, while maintaining clear operation for part of the time for the 802.11b WLAN.

COEXISTENCE RESULTS

The mechanisms for coexistence in 802.15.3 will enable other WPANs and WLANs to operate in close proximity to 802.15.3 piconets. This section presents some of the calculated results for 802.15.3 piconets operating in proximity with other wireless networks.

Assumptions for coexistence simulations

Although it is possible to do an in-depth simulation and modeling of the 802.15.3 MAC and 2.4 GHz PHY combination, the results would depend heavily on the type and quantity of traffic assumed in each of the wireless networks. Instead, 802.15.3 used a simple PHY layer model, proposed in IEEE Std 802.15.2 [B15], to gauge the impact of 802.15.3 piconets sharing operational areas with 802.15.1 piconets and 802.11b networks.

The calculations rely on IEEE Std 802.15.1 [B14], IEEE Std 802.11 [B12], IEEE Std 802.11b [B13], and IEEE Std 802.15.2 [B15]. For the calculations used to determine the level of coexistence, the following assumptions have been made:

a) The sensitivity for each of the receivers is the reference sensitivity given in each of the standards, i.e.,

  1) −70 dBm for 802.15.1,

  2) −76 dBm for 802.11b 11 Mb/s CCK, and

  3) −75 dBm for 802.15.3 22 Mb/s DQPSK.

b) The received power at the desired receiver is 10 dB above the receiver sensitivity. The level of 10 dB was selected because a 10-dB margin results in 10% frame error ratio (FER) in a Raleigh fading channel. Reliable communications without interference would require at least this margin. The distance between the desired transmitter and receiver is not directly specified by this requirement; instead, the transmitter power and the channel model listed below would be used to determine the resulting distance.

c) The transmitter power for each of the protocols is

1) 0 dBm for 802.15.1,

2) +14 dBm for 802.11b, and

3) +8 dBm for 802.15.3.

The path loss model is one that was proposed for 802.11 and used by 802.15.2:

$$d = 10^{\frac{(P_t - P_r - 40.2)}{20}} \quad \text{for } d < 8 \text{ m}$$

$$d = 8 \times 10^{\frac{(P_t - P_r - 58.5)}{33}} \quad \text{for } d > 8 \text{ m}$$

The path loss is plotted versus the range in Figure 9–2.

d) The receiver bandwidths are based on the requirements in the standard:

1) 1 MHz for 802.15.1,

2) 22 MHz for 802.11b, and

3) 15 MHz for 802.15.3.

e) The transmitter spectral masks are the maximum allowed in the standards. This is a very pessimistic assumption because the transmitter sideband spectrum will generally be significantly lower than the spectral mask over most of the frequencies. There are usually only narrow peaks that come close to the required limits. The subclauses that define the transmitter spectral mask for the three standards are as follows:

1) Subclause 7.2.3.1 for 802.15.1,

2) Subclause 18.4.7.3 for 802.11b, and

3) Subclause 11.5.3 for 802.15.3.

f) The energy from the interfering signal affects the desired signal in a manner equivalent to additive white Gaussian noise (AWGN) in the same bandwidth.

g) The 802.15.3 piconet operates with the 802.11b coexistence channel plan as described in subclause 11.2.3 of IEEE Std 802.15.3.

For all of the results in this section, the FER calculations assume that one of the two systems is constantly transmitting, while the other one is constantly receiving. If both systems are either both transmitting or both receiving, then the FER will be essentially unaffected by the presence of the other system. For random transmission and reception, this means that the actual FER will only be about one-half of the values presented in the figures. In addition, the MAC protocol and the traffic patterns will affect the actual throughput that is possible with the collocated networks.

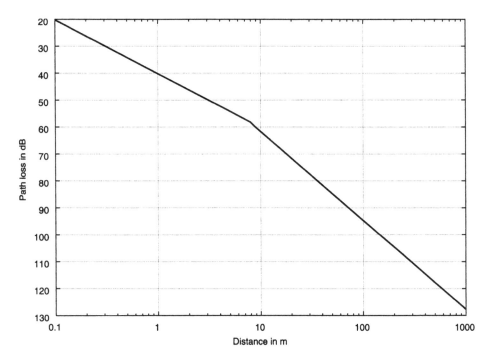

Figure 9–2: Path loss versus distance for empirical model

BER calculations

The BER calculations for these simulations are based on the formulas suggested in IEEE Std 802.15.2 [B15]. For 802.11b and 802.15.3, the basic BER calculation depends on the theoretical formula for the BER of QPSK modulation, which in turn depends on the value of the Q function. The Q function is defined as the area under the tail of the Gaussian probability density function with zero mean and unit variance.

$$Q(x) = \frac{1}{\sqrt{2\pi}} \int_x^\infty e^{-\left(\frac{t^2}{2}\right)} dt$$

Although the Q function is not commonly found in software packages, it can be calculated from the complementary error function, erfc, which is more common and can be found in spreadsheets.

$$Q(x) = 0.5 \text{ erfc}\left(\frac{x}{\sqrt{2}}\right)$$

If neither the Q function nor the erfc are available, the Q function may be approximated using the following fifth-order approximation for values of x greater than 1.

$$Q(x) = \left(\frac{1}{\sqrt{2\pi}}\right) e^{-\left(\frac{t^2}{2}\right)} \times \left(\frac{8 + 9x^2 + x^4}{15x + 10x^3 + x^5}\right)$$

The probability of error in the various modulations can the be calculated from the following equations (see [B15] for more details on the derivations):

> **Notice** The equations here were derived assuming CCK could be modeled as a block code. This assumption is invalid for small values of the SNR and so these equations will give poor results for low SNR values.

$$\text{BER}_{22 \text{ Mb/s } 802.15.3} = Q(\sqrt{SNR})$$

$$\text{BER}_{1 \text{ Mb/s } 802.11} = Q(\sqrt{11 \times SNR})$$

$$\text{BER}_{2 \text{ Mb/s } 802.11} = Q(\sqrt{5.5 \times SNR})$$

$$\text{BER}_{5.5 \text{ Mb/s } 802.11} = \frac{8}{15}[14 \times Q(\sqrt{8 \times SNR}) + Q(\sqrt{16 \times SNR})]$$

$$BER_{11 \text{ Mb/s } 802.11} = \frac{128}{255}[24 \times Q(\sqrt{4 \times SNR}) + 16 \times Q(\sqrt{6 \times SNR})$$
$$+ 174 \times Q(\sqrt{8 \times SNR}) + 16 \times Q(\sqrt{10 \times SNR})$$
$$+ 24 \times Q(\sqrt{12 \times SNR}) + Q(\sqrt{16 \times SNR})]$$

Determining the BER for 802.15.1 is a little more difficult. The equation for the BER of orthogonal FSK is given by

$$BER_{802.15.1} = 0.5 \times e^{(0.5 \times SNR)}$$

Although this is a very simple equation, 802.15.1 uses non-orthogonal FSK, which does not have a closed-form solution. However, this approximation is sufficient for the present analysis.

Figure 9–3 illustrates the relationship between BER and SNR for 802.11b, 802.15.3 base rate, and 802.15.1. In the figure, the BER values for the 5.5 Mb/s and 11 Mb/s data rates of 802.11b are clipped at a probability of 0.5 because the BER equations for these data rates are inaccurate for low SNR values. As the SNR value decreases, the BER should approach a value of 0.5. However, the equations for the BER of 5.5 Mb/s and 11 Mb/s will approach values of 4 and 64 as the SNR approaches zero.[44] However, for high BER values, i.e., greater than 0.01, the FER is essentially 1 for frame sizes of larger than 256 octets, and so the inaccuracy in the BER calculation would not affect the FER values.

802.11b and 802.15.3

802.11b networks have become much more common in recent years, and so coexistence with these types of networks is important for 802.15.3 applications. Despite the relative success of 802.11b, it has not yet achieved widespread penetration into peoples' homes, which is the target market of 802.15.3 networks. In addition, although it is common for office deployments of 802.11b to use more than one of the available channels, in a typical home, there will be only one

---

[44] These values can be calculated by noting that the Q function is equal to 0.5 when its argument is zero.

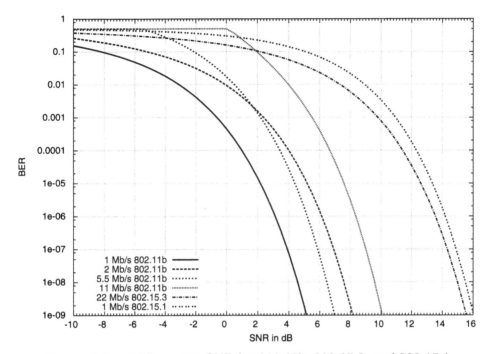

Figure 9–3:   BER versus SNR for 802.11b, 802.15.3, and 802.15.1

802.11b network, and so the other two channels will be available for the 802.15.3 WPAN to use. Thus, in many cases, interference from other sources, e.g., 2.4 GHz phones or microwave ovens, will be more important than will the interference from 802.11b WLANs.

If an 802.15.3 piconet operates in the same area as an 802.11b WLAN, the PNC will notice the reduced throughput and will search for the quietest channel. In addition, if it detects the presence of an 802.11b network, it will use the 802.11b coexistence channel plan to avoid overlapping with the WLAN. In addition, the PNC will rate that channel as the worst to avoid choosing the occupied channel.

The current IEEE Std 802.11b does not provide the ability for an AP to request neighbor piconet status, and it does not prohibit the AP from doing it either. As IEEE Std 802.15.3 is very new, there are not any products on the market that provide this capability. However, it is likely that vendors that want to serve both markets will develop a dual-mode AP that enables the two networks to share the channel capacity. One of the applications for this is a home where the consumer electronic devices, e.g., televisions, CD players, camcorders, and so on, are networked with 802.15.3, but the network also needs to support an 802.11b laptop for Internet access. A dual-mode AP could allocate time for the 802.11 contention period and use contention-free period to set aside time for the 802.15.3 WPAN. One of the issues with this approach is that the end of the contention period in 802.11 is not fixed in time. Instead, it ends when the AP gets a chance to send a beacon after the end of a transmission by a STA. In this case, it would be possible for the 802.11 network to continue transmitting past the time allocated for it by the PNC.

Using the modeling assumptions stated in "Assumptions for coexistence simulations" on page 278, the degradation of the FER of an 802.15.3 piconet in the presence of an 802.11b WLAN is illustrated in Figure 9–4. In the graph, the adjacent channel is 25 MHz away, whereas the alternate channel is separated by 50 MHz.

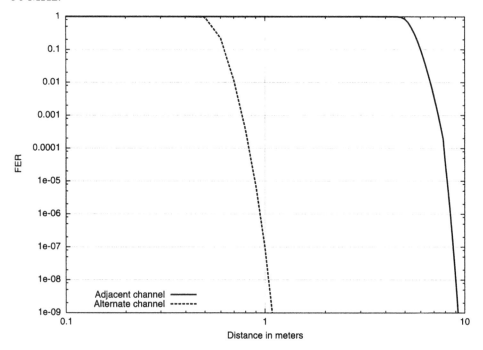

Figure 9–4: FER results for 802.15.3 with 802.11b as the interferer

The adjacent and alternate channels for 802.15.3 as the receiver with 802.11b as the interferer are given in Table 9–3.

Table 9–3: Adjacent and alternate channels for 802.15.3 as receiver

802.15.3 channel	Adjacent 802.11b channel	Alternate 802.11b channel
1	6	11
3	1, 11	none
5	6	1

As Figure 9–4 shows, there is almost no effect from the 802.11b STA when it is in the alternate channel of the 802.15.3 DEV. However, when the 802.11b STA is in the adjacent channel, the 802.15.3 DEV shows a noticeable degradation in performance (i.e., an FER of >3%) when the separation is less than 8 m. The 802.11b STA causes a degradation in the 802.15.3 DEV's throughput due to its larger occupied bandwidth and higher transmit power.

The 802.15.3 DEV will degrade the FER of 802.11b STA as well. This is shown in in Figure 9–5 for the 802.15.3 DEV in the adjacent channel and in Figure 9–6 for the 802.15.3 DEV in the alternate channel. When the 802.15.3 DEV is in the adjacent channel, the 802.11b STA does not experience a noticeable degradation in the FER for any of the data rates until the range is reduced to less than 0.8 m. In the case when the 802.15.3 DEV is in the alternate channel, this is reduced to a 0.4 m range.

The 802.15.3 DEV has very low impact on the 802.11b STA for two reasons. The first reason is that the 802.15.3 DEV has a relatively narrow spectrum, and so it has very little transmit energy in the 802.11b channel. The second reason is that the 802.15.3 DEV uses a lower transmit power than does the 802.11b STA. The results for the adjacent and alternate channels are much closer in this case due to the more stringent spectral mask of 802.15.3. The adjacent and alternate channel numbers for 802.11b as the receiver with 802.15.3 as the interferer are given in Table 9–4.

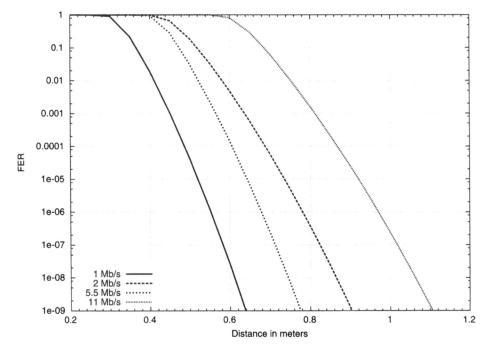

Figure 9–5: FER results for 802.11b with 802.15.3 as the interferer in the adjacent channel

Table 9–4: Adjacent and alternate channels for 802.11b as receiver

802.11b channel	Adjacent 802.15.3 channel	Alternate 802.15.3 channel
1	3	5
6	1, 5	none
11	3	1

If the 802.15.3 DEV and the 802.11b STA are in the same channel, then the degradation is much more significant. The FER for this case is graphed in Figure 9–7. For both 802.11 and 802.15.3, the performance of the networks are dramatically reduced. In the case of the 802.11 STA, the performance begins to degrade when the distance is less than 60 m for 11 Mb/s and less than 50 m for the

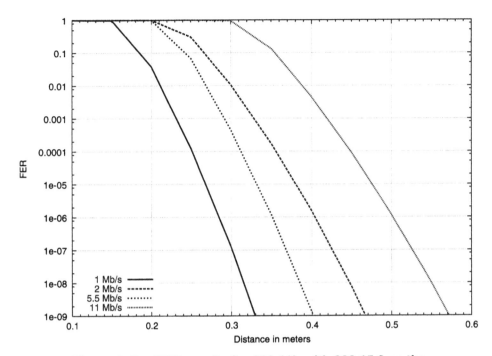

Figure 9–6:   FER results for 802.11b with 802.15.3 as the interferer in the alternate channel

1 Mb/s data rate. The 802.15.3 DEV, on the other hand, has its throughput affected up to a separation of almost 80 m.

The results for the co-channel interference illustrate the value provided by 802.15.3's ability to automatically change channels and to allocate time in the current channel to other networks. Although it is possible to change the channel for an 802.11b WLAN, it requires user intervention to change the setting on the AP. In 802.15.3, the PNC automatically changes the channel whenever the conditions require it. In addition, because all of the DEVs in the WPAN are notified of the upcoming change, the channel switch will have a minimal effect on the current connections.

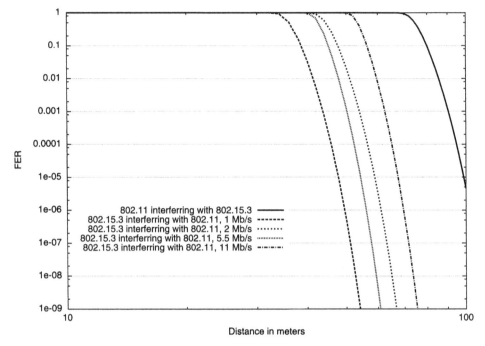

**Figure 9–7: FER results for 802.11b and 802.15.3
co-channel interference**

802.15.1 and 802.11 FHSS overlapping with 802.15.3

802.15.1 and 802.11 FHSS systems have similar characteristics in terms of the coexistence performance with respect to 802.15.3. In practice, the protocol differences will give somewhat different results. In particular, the faster hopping time of 802.15.1 can cause more problems with long data frames, especially ones sent at the lower data rates. In addition, the TDMA method used in 802.15.1 causes more problems with the CSMA/CA-based access method used in 802.11 than with the TDMA method used in 802.15.3. The CSMA/CA STA will lose frames when the TDMA DEV overlaps transmissions. In addition, the CSMA/CA STA will back off on the TDMA DEV's transmissions, reducing the throughput. The backoff window is also increased when the frame is dropped due to the collision. For the present analysis, the effects of the protocol are ignored. As most of the 2.4 GHz cordless phones are also FHSS systems, the results here are similar to those that would be obtained for interference with these systems.

For a frequency hopping system, the effect on the 802.15.3 FER will depend on the hop frequency. When the FHSS system is on a frequency that is far away from the 802.15.3 channel, the effect on either system will be low. On the other hand, when the hop frequency overlaps with the 802.15.3 channel, the effect is much worse because there is no filtering in either the transmitter or the receiver to reduce the interfering power. The overall FER can be calculated as the average of the FERs for each of the hop frequencies because the FHSS systems will visit each of the frequencies an equal number of times.[45]

The average FER with 802.15.1 as the receiver and 802.15.3 as the interferer ("802.15.1 receiving") and with 802.15.3 as the receiver and 802.15.1 as the interferer ("802.15.3 receiving") is shown in Figure 9–8.

The FER curves are flat over a large separation distance. In this region, the systems do not affect each other when the hop frequency is outside of the 802.15.3 channel. On the other hand, when the hop frequency falls inside of the 802.15.3 channel bandwidth, both systems experience 100% lost frames for these distances. As the FHSS systems use 79 1 MHz wide channels, these overlap with the 15 MHz wide 802.15.3 channel on average 15 out of 79 times or about 19% of the time. This creates a 19% FER, as shown in the figure. Once the two systems have been separated by sufficient distance, the FER due to co-channel interference begins to drop and so the overall FER begins to fall as well.

The performance of the 802.15.3 system can be improved using either a notch filter or an equalizer that can remove some of the power in the interfering signal. If the FHSS system is collocated with the 802.15.3 system, then the frequency of the notch filter will be known. If the systems are not collocated, it is possible for the 802.15.3 system to learn the hopping pattern and predict the required location of the notch filter. Finally, if the interferer starts prior to the 802.15.3 frame, then the equalizer in the modem will tend to reduce the effect of the narrowband interferer.

---

[45] This is not true with 802.15.1 because the hop sequence is selected from a poorly-designed sequence generator that does not generate random sequences. However, the assumption that hop sequence for 802.15.1 is pseudo-random is sufficient for the present analysis.

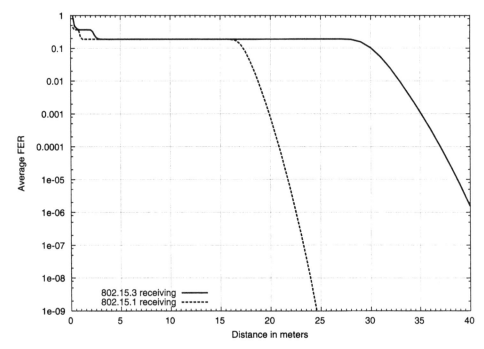

Figure 9–8: Average FER results for 802.15.1 and 802.15.3

Summary

IEEE Std 802.15.3 is the first IEEE standard specifically designed to support wireless multimedia applications in low-cost, portable devices. Because of this, the standard is poised to become widely adopted in consumer electronic devices that need multimedia connectivity. The standard supports applications that require both high throughput and good QoS, e.g., wireless distribution of video and high-quality audio. The standard was also designed to provide high throughput with a latency that is low and predictable. These characteristics support applications such as telephony, video conferencing, gaming, remote control, and channel surfing. Although the standard has focused on support for wireless multimedia, asynchronous traffic is supported as well, providing high throughput for that traffic when the channel time is not required for higher priority streams.

During the development of the standard, the 802.15.3 Task Group kept its focus on enabling low-cost, portable implementations. Both the MAC and PHY were specified to allow compliant devices with minimal functionality for low-cost implementations, while providing the ability to create full-featured implementations that maintain compatibility. Significant effort went into the development and analysis of the power-saving modes that are required for portable applications. The power-save modes in the standard range from a very simple, unsynchronized deep-sleep mode to a very efficient mode that provides predictable latency and low-rate, low-power QoS communication.

The 802.15.3 MAC architecture is simple, but scalable, providing the structure necessary to develop complex applications. The most basic compliant entity is the DEV, which only needs to support CSMA/CA for data transfer. A DEV can also be made very complex, supporting the sourcing and sinking of multiple streams with efficient power savings, device discovery, and secure communications. Likewise, the simplest PNC only needs to support sending the beacon, associating a limited number of DEVs and maintaining a heartbeat counter for each of the associated DEVs. A PNC can also be a more complex implementation that is able to monitor the piconet conditions, instantly adapt the allocations to handle changes in the environment or in data, and control more than one piconet. The standard has already planned for future growth by reserving fields and numbers for new ammendments while also providing methods for implementors to differentiate their products without collisions that create interoperability problems.

The combination of CSMA/CA and TDMA in the superframe supports the efficient transfer of commands and small amounts of data, while providing the guaranteed allocations necessary for QoS. The CAP is easy for the PNC to allocate and provides a simple contention method for DEVs to send commands or even small amounts of data. However, the bulk of the data in the superframe is carried in channel time allocations using TDMA as the access method. Using TDMA supports the requirements for QoS while providing throughput efficiencies that are 30% to 100% greater than using a CSMA/CA method. The combination of the 802.15.3 MAC and PHY allow piconet to achieve a throughput in excess of 80% of the actual data rate when all of the MAC and PHY overhead is included. Providing high throughput and guaranteed allocations are only two of the requirements for supporting QoS. IEEE Std 802.15.3 also allows a DEV to request allocations that control the latency of its data transmissions. The channel time allocation request is used by the DEV to request the spacing between allocations, either multiples of a superframe, in the case of sub-rate allocations, or multiple times in a superframe, in the case of super-rate allocations.

While the MAC sublayer provides the basis for delivering multimedia connections, it is the PHY that makes it wireless. The 2.4 GHz PHY is based on proven technology that has decades of experience in various implementations. The technology for 2.4 GHz unlicensed applications has matured to the point that single-chip radios are commonplace and complete radio solutions are both small in size as well as being very low cost. The supporting parts, e.g., filters, baluns, couplers, power amplifiers, etc., are also widely available at very low-cost points and in a variety of form factors. The 2.4 GHz band has the advantage of near world-wide availability for unlicensed applications. The 2.4 GHz band also has better range characteristics than either 5 GHz or UWB allocations because the of the lower frequency and, in the case of UWB, much higher transmit power. Single carrier modulation was selected for the PHY because it can be easily implemented in very few gates. It also has very low required signal to noise ratios, extending the range for essentially no additional cost or power.

Because of the good propagation characteristics and wide availability of its frequency allocation, the 2.4 GHz band has attracted many different systems, including IEEE Std 802.11b, IEEE Std 802.11g, IEEE Std 802.15.1, and cordless telephones. The 802.15.3 standard uses both MAC and PHY techniques to enhance the coexistence of 802.15.3 piconets with these types of interferers. MAC techniques for coexistence include passive scanning, dependent piconets, channel status request, remote channel scan, and dynamic channel selection. The PHY improves coexistence with the choice of operating frequencies, reduced transmit and receive spectral mask, link quality, RSSI and transmit power control. These techniques not only reduce the impact of 802.15.3 on existing wireless devices, they also improve the stability and throughput of 802.15.3 piconets.

IEEE Std 802.15.3 is uniquely capable of supporting high-throughput wireless multimedia connections between low-cost, low-power portable devices. Applications that require good QoS are well supported by the standard. Although the MAC provides for simple, interoperable devices, it also supports interoperability for more full-featured implementations and applications. The flexibility of the standard makes it easy for a DEV to request and receive the exact type of channel time allocations needed to support the DEV's applications. The PHY portion of the standard builds on many years of radio design and leverages the popularity of current 2.4 GHz solutions in order to keep the cost of the PHY layer very low. The PHY allows simple low-cost implementations while also defining high-rate modes that allow for high-speed data connections. Coexistence with existing networks was taken into consideration in the development of the standard and so the performance of 802.15.3 piconets in the presence of interferers is very good. The many advantages of 802.15.3 will help to make it the default choice for low-power, low-cost, wireless multimedia connectivity in consumer electronic devices.

References

[B1] 47 CFR Part 15, Radio Frequency Devices, Federal Communications Commission. http://www.fcc.gov/oet/info/rules/Welcome.html.

[B2] ARIB STD-T66, "Second generation low-power data communication system/wirelss LAN system," Association of Radio Industries and Businesses, http://www.arib.or.jp.

[B3] Bailey, D., Singer, A. and Whyte, W., "NTRUEncrypt Security Suite," submission to IEEE Task Group 802.15.3, document number 02362r0P802-15_TG3-NTRUEncrypt-Security-Suite.pdf, http://grouper.ieee.org/groups/802/15/pub/2003/Sep02/.

[B4] Bailey, D., Singer, A. and Whyte, W, "RSA Security Suite," submission to IEEE Task Group 802.15.3, document number 02363r0P802-15_TG3-RSA-Security-Suite.pdf, http://grouper.ieee.org/groups/802/15/pub/2003/Sep02/.

[B5] Batra, Anuj, et. al., "Physical Layer Submission to 802.15 Task Group 3a: Time-Frequency Interleaved Orthogonal Frequency Division Multiplexing (TFI-OFDM)," submission to IEEE Task Group 802.15.3, document number 03141r3P802-15_TG3a-TI-CFP-Presentation.ppt, http://grouper.ieee.org/groups/802/15/pub/2003/May03/.

[B6] ETSI EN 300 328 V1.4.1 (2003-04), "Electromagnetic compatibility and Radio spectrum Matters (ERM); Wideband Transmission systems; Data transmission equipment operating in the 2,4 GHz ISM band and using spread spectrum modulation techniques; Harmonized EN covering essential requirements under article 3.2 of the R&TTE Directive," European Telecommunications Standards Institute, http://www.etsi.org.

[B7] ETSI ETS 300 826 ed.1 (1997-11), "Electromagnetic compatibility and Radio spectrum Matters (ERM); ElectroMagnetic Compatibility (EMC) standard for 2,4 GHz wideband transmission systems and HIgh PErformance Radio Local Area Network (HIPERLAN) equipment," European Telecommunications Standards Institute, http://www.etsi.org.

[B8] Hawbaker, D. A., "Indoor wide band radio wave propagation measurements and models at 1.3 and 4.0 GHz," Masters thesis, Electrical Engineering, Virginia Polytechnic Institute State University, May 1991.

[B9] Huang, C.-C., and Khayata, R., "Delay spreads and channel dynamics measurements at ISM bands," *SUPERCOMM/ICC '92 Digest*, Chicago, IL, pp. 1222–1226, June 1992.

[B10] IEEE Std 802.1D™, 1998 Edition, IEEE Standard for Information technology—Telecommunications and information exchange between systems—Local and metropolitan area networks—Common specifications—Part 3: Media Access Control (MAC) Bridges.[46]

[B11] IEEE Std 802.2™, 1998 Edition (R2003), IEEE Standard for Information technology—Telecommunications and information exchange between systems—Local and metropolitan area networks—Specific requirements—Part 2: Logical Link Control.

[B12] IEEE Std 802.11™, 1999 Edition (R2003), IEEE Standard for Information technology—Telecommunications and information exchange between systems—Local and metropolitan area networks—Specific requirements—Part 11: Wireless LAN Medium Access Control (MAC) and Physical Layer (PHY) Specifications.

[B13] IEEE Std 802.11b™-1999 (R2003), Supplement to IEEE Standards for Information technology—Telecommunications and information exchange between systems—Local and metropolitan area networks—Specific requirements—Part 11: Wireless LAN Medium Access Control (MAC) and Physical Layer (PHY) Specifications: Higher-Speed Physical Layer Extension in the 2.4 GHz Band.

[B14] IEEE Std 802.15.1™-2002, IEEE Standard for Information technology—Telecommunications and information exchange between systems—Local and metropolitan area networks—Specific requirements—Part 15.1: Wireless Medium Access Control (MAC) and Physical Layer (PHY) Specifications for Wireless Personal Area Networks (WPANs).

---

[46] IEEE publications are available from the Institute of Electrical and Electronics Engineers, Inc., 445 Hoes Lane, P. O. Box 1331, Piscataway, NJ 08855-1331, USA. IEEE publications can be ordered on-line from the IEEE Standards website: http://www.standards.ieee.org.

[B15] IEEE Std 802.15.2™-2003, IEEE Recommended Practice for Information technology—Telecommunications and information exchange between systems—Local and metropolitan area networks—Specific requirements—Part 15.2: Coexistence of Wireless Personal Area Networks with Other Wireless Devices Operating in Unlicensed Frequency Bands.

[B16] IEEE 802.15.3™-2003, IEEE Standard for Information technology—Telecommunications and information exchange between systems—Local and metropolitan area networks—Specific requirements—Part 15.3: Wireless Medium Access Control (MAC) and Physical Layer (PHY) Specifications for High Rate Wireless Personal Area Networks (WPANs).

[B17] IEEE Std 1394a™-2000, IEEE Standard for a High-Performance Serial Bus—Amendment 1.

[B18] IEEE Std 1394b-2002, IEEE Standard for a High-Performance Serial Bus—Amendment 2.

[B19] Intel UWB Database, Ultra Lab, University of Southern California. http://impulse.usc.edu.

[B20] MacLellan, J., Lam, S., and Lee, X., "Residential indoor RF channel characterization," *Proceedings 1993 IEEE Vehicular Technology Conference*, Secaucus, NJ, pp. 210–213, 1993.

[B21] Merrit, R, "Group hopes to leapfrog 802.11 for wireless video," *Electronic Engineering Times*, May 3, 2003, http://www.eet.com/story/OEG20030502S0060.

[B22] NIST FIPS Pub 197: Advanced Encryption Standard (AES), Federal Information Processing Standards Publication 197, U.S. Department of Commerce/NIST, November 26, 2001.

[B23] Nobles, P., Ashworth, D., and Halsall, F., "Propagation measurements in an indoor radio environment at 2, 5 and 17 GHz," *IEEE Colloquium on High Bit Rate UHS/SHF Channel Sounders—Technology and Measurement*, pp. 4/1–4/6, 1993.

[B24] Rappaport, T. S., and Hawbaker, D. A., "Wide-band microwave propagation parameters using circular and linear polarized antennas for indoor wireless channels," *IEEE Transactions in Communications*, vol. 40, no. 2, pp. 240–245, Feb. 1992.

[B25] RSS-210, "Specification Low-Power Licence-Exempt Radiocommunication Devices (All Frequency Bands)," Industry Canada, http://www.strategis.ic.ga.ca/engdoc/main.html.

[B26] Stein, J. C., *Indoor Radio WLAN Performance Part II: Range Performance in a Dense Office Environment*, Harris (now Conexant) Semiconductor web page, http://www.intersil.com/data/wp/WP0546.pdf.

[B27] Struik, R. and Rasor, G., "Mandatory ECC Security Algorithm Suite," submission to IEEE Task Group 802.15.3, document number 02200r3P802-15_TG3-Mandatory-ECC-Security-Algorithm-Suite.doc, http://grouper.ieee.org/groups/802/15/pub/2003/Sep02/.

[B28] TIA/EIA-95-B: Mobile Station-Base Station Compatibility Standard for Wideband Spread Spectrum Cellular Systems (ANSI/TIA/EIA-95-B-99), Committee TR-45.5, February 1, 1999.

[B29] "Title 47 of the Code of Federal Regulations, Part 15," http://www.fcc.gov/oet/info/rules/.

[B30] Weaver, D. K., Jr., "A third method of generation and detection of single-sideband signals," *Proceedings of the IRE*, vol. 44, no. 12, pp. 1703–1705, June 1956.

[B31] Yoshida, J., "Bluetooth gropes for identity," *Electronic Engineering Times*, page 1, June 23, 2003.

[B32] Zyren, J., Enders, E. and Edmondson T., "802.11g Starts Answering WLAN Range Questions," *CommDesign.com*, January 14, 2003, http://www.commsdesign.com/story/OEG20030114S0008.

Glossary

ACTIVE mode: A power management mode in which the device (DEV) is in AWAKE state for every superframe to receive the beacon. The DEV is also in AWAKE state during every superframe for the contention access period (CAP) and any channel time allocations (CTAs) for which it is either the source or the destination.

ad hoc network: A network that is created without the intervention of the user. Ad-hoc networks are typically formed as they are needed and dissolved when the desired communication is complete.

association: In the context of IEEE Std 802.15.3, this is the process by which a device (DEV) joins an existing piconet. The assocaition process allows the piconet controller (PNC) to refuse access to DEVs if its resources are limited. It is also used to assign the 8-bit device identifier (DEVID).

asynchronous power save (APS) mode: A power-save mode in which the device (DEV) is in SLEEP state and wakes only to send a frame to the piconet controller (PNC) to maintain its associated status. Unlike device synchronized power save (DSPS) and piconet synchronized power save (PSPS) modes, a DEV in asynchronous power save (APS) mode is free to wake up in any superframe of its choosing to send a frame to maintain its association with the PNC and does not have to be in AWAKE state for any specific beacons.

AWAKE state: The condition of the device (DEV) when it is either transmitting or receiving.

channel time allocation (CTA): A specified duration of time assigned by the piconet controller (PNC) to a specific source device (DEV) during the superframe for its exclusive use. During this time, no other DEV in the piconet is allowed to begin a frame exchange. The destination can be either a single DEV, a group of DEVs (multicast), or all of the DEVs in the piconet (broadcast).

channel time allocation period (CTAP): The period of time in the 802.15.3 superframe set aside for channel time allocations. In general, devices (DEVs) do not contend for channel time during the CTAP, rather it is assigned to a DEV by the piconet controller (PNC) in response to a request from that DEV. Channel time allocations use time division multiple access (TDMA), with the exception of association management channel time allocations (MCTAs) and open MCTAs, which use slotted aloha.

child piconet: A dependent piconet in which the piconet controller (PNC) is a full member of the parent piconet.

child piconet controller (PNC): A PNC that is a full member of another piconet, called the parent piconet, and synchronizes its piconet so that its communications occur only during a channel time allocation (CTA) granted to it by the parent PNC.

clear channel assessment (CCA): The process by which a radio determines that the wireless medium is clear. A radio performs CCA for 802.15.3 by enabling its reciever and searching either for the preamble or by determining that the power in the channel exceeds a specified level.

contention access period (CAP): The period of time in the 802.15.3 superframe set aside for devices (DEVs) to use collision sense multiple access with collision avoidance (CSMA/CA) to gain access to the medium. The CAP can be used to send data and commands.

dependent piconet: A piconet that operates during a channel time allocation in the superframe of the parent piconet. The parent piconet controller (PNC) assigns the channel time for the dependent piconet in response to a request from the dependent PNC. There are two types of dependent piconets defined in the IEEE Std 802.15.3, child and neighbor piconets.

dependent piconet controller (PNC): A PNC that has synchronized piconet timing to another piconet, called the parent piconet, so that the communications occur only during a channel time allocation (CTA) granted to it by the parent PNC. The dependent PNC can be either a child PNC or neighbor PNC depending on its membership status in the parent piconet.

device (DEV): Any of the participants in an IEEE Std 802.15.3 piconet. The piconet controller (PNC) maintains a DEV personality in the piconet even as it operates as the PNC.

device synchronized power save (DSPS): A power save mode in which devices (DEVs) wake up periodically to listen to beacons. DSPS mode differs from piconet synchronized power save (PSPS) mode in that the DEV specifies the sleep interval for DSPS mode, while the piconet controller (PNC) determines it for PSPS mode.

disassociation: The process by which a device (DEV) either leaves the piconet or is removed from the piconet by the PNC.

extended beacon: A method for that allows the beacon transmissions to consist of relatively small frames for reliability. The extended beacon is composed of a beacon frame followed by one or more broadcast Announce commands and the appropriate interframe spacings.

fragmentation: The process by which the protocol splits up a larger data unit into smaller ones, adding information that will allow the receiving entity to reconstruct the original data unit.

frame: The basic on-air data unit for IEEE Std 802.15.3. It is composed of the PHY preamble, PHY header, MAC header, MAC header protection, data, security information (for secure frames), and frame check sequence.

integrity code (IC): A string of bits generated with a symmetric encrypting key that is typically appended to data in order to provide data integrity and source authentication similar to a digital signature.

MAC protocol data unit (MPDU): A block of data exchanged between the MAC and the PHY either to be sent by the PHY or because it was received by the PHY.

MAC service data unit (MSDU): A block of data exchanged between the MAC and the frame convergence sublayer (FCSL) either to be sent over the wireless medium by the PHY or because it was received by the MAC from the PHY.

neighbor piconet: A dependent piconet in which the piconet controller (PNC) is not a full member of the parent piconet.

neighbor piconet controller (PNC): A PNC that has associated with another piconet, called the parent piconet, but which is not a full member of the parent piconet. The neighbor PNC synchronizes its wireless network with the parent piconet in such a way as to guarantee that transmissions from the neighbor piconet occur only during a channel time allocation (CTA) granted to it by the parent PNC.

parent piconet: A piconet that provides channel time allocations for another piconet, referred to as a dependent piconet, to allow it to operate in the same PHY channel.

piconet: The basic configuration of an IEEE Std 802.15.3 network. A piconet is formed when the piconet controller (PNC) begins sending beacons. All piconets have a PNC and at least one device (DEV) (which is just the PNC's DEV personality). The smallest non-trivial instantiation of an 802.15.3 piconet consists of the PNC and two DEVs, one of which is the PNC's DEV personality and one DEV that is distinct from the PNC.

piconet controller (PNC): A device (DEV) that is in charge of allocating channel time and allowing access to the piconet via association. The PNC can also act as the security coordinator for the piconet, performing authentication and key management for the members of the piconet.

piconet synchronized power save (PSPS): A power-save mode in which the device (DEV) is required to be in AWAKE state for beacons specified by the piconet controller (PNC). In PSPS mode, the PNC determines the periodicity (if any) of the wake beacons, as opposed to device synchronized power save (DSPS) mode where this is determined by the DEV.

power management (PM) mode: In IEEE Std 802.15.3, this is the current status of a device (DEV) with respect its desire to save power. 802.15.3 defines four PM modes: ACTIVE, device synchronized power save (DSPS), piconet synchronized power save (PSPS), and asynchronous power save (APS).

power-save (PS) mode: In IEEE Std 802.15.3, this refers to a mode in which the device (DEV) has reduced its activity in order to save power. 802.15.3 defines three power-save modes: device synchronized power save (DSPS), piconet synchronized power save (PSPS), and asynchronous power save (APS).

quality of service (QoS): A measure of the performance of a network with respect to certain basic performance criteria, including latency, reliability (low downtime), and error performance. This metric can be used to compare the ability of different types of networks with respect to delivering time-critical information.

secure frame: A command or data frame that has been protected with an integrity code and, in the case of data frames, encrypted to prevent eavesdropping.

secure membership: A device (DEV) achieves secure membership in the piconet when it has recieved a management key from the piconet controller (PNC). Once the DEV has the management key, it can also get the data key to encrypt and decrypt data as well as to authenticate commands.

secure piconet: A piconet in which the beacon and most command frames are protected with an integrity check and data frames can be encrypted with a common symmetric key pair.

SLEEP state: The condition where the device (DEV) is neither transmitting nor receiving.

stream: A unidirectional data connection between a source device (DEV) and one or more destination DEVs. Streams can be used for both isochronous and asynchronous data.

sub-rate allocation: A channel time allocation that occurs only once every n superframes ($n > 1$) where n is the CTA Rate Factor sent in the Channel Time Request command. For example, a CTA Rate Factor of 4 would result in an allocation that occurs in every fourth superframe. Sub-rate allocations are used for low-data rate streams or to maintain connectivity between sleeping devices (DEVs) in device synchronized power save (DSPS) mode.

super-rate allocation: A channel time allocation that occurs at least once in every superframe.

superframe: The basic timing period for an IEEE Std 802.15.3 piconet. The superframe is composed of the beacon, the channel time allocation period, and optionally the contention access period.

synchronized power save (SPS): A power-save mode in which devices (DEVs) wake up regularly for the same beacon, called wake beacons, which occur at regular intervals, e.g., every eight superframes. There are two types of SPS modes defined in IEEE Std 802.15.3: device synchronized power save (DSPS) and piconet synchronized power save (PSPS).

wake beacon: The beacon during which a synchronized power save (SPS) device (DEV), either piconet synchronized power save (PSPS) or device synchronized power save (DSPS), will be in the AWAKE state. All of the members of a given SPS set are required to be in the AWAKE state for the wake beacon.

Index

Numerics

802.11 3, 5, 6, 30, 31, 33, 45, 49, 104, 166, 203–208, 223, 230, 273, 274, 278, 279, 282–289

A

access control list (ACL) 190, 92, 194, 199, 200

acknowledgement (ACK) 60, 78, 81–86, 160, 170, 212

    delayed (Dly-ACK) 81–99, 119, 208, 212

    Dly-ACK negotiation 84–86

    immediate (Imm-ACK) 47–49, 61, 75, 80–83, 85, 94, 99, 119, 167, 208, 212

    no (no-ACK) 53, 55, 75, 81, 94–98, 149–150, 208

    policy 75, 80–86, 119

    request 81, 83, 85, 86

ACTIVE 40, 42, 43, 102, 103, 104, 106, 111–114, 118–129, 146, 150, 169, 179, 263

advanced encryption standard (AES) 189, 190, 195–197

advanced encryption systems (AES) 279

antenna gain 30, 205, 213, 231

association 26, 29, 32–35, 37, 56, 57, 58, 60, 61, 66, 73–78, 80, 81, 103, 145, 155, 159, 160, 168–173, 180, 183, 194

Association Request 168

association timeout period (ATP) 35, 41, 75, 76, 79, 80, 91, 103, 105, 127, 131, 146, 193

asynchronous power save (APS) 40, 41, 42, 102–107, 113, 114, 126–129

Authentication 279

authentication   6, 11, 12, 24, 70, 76, 187–195, 199

    mutual 192

authorization 70, 73, 187–189, 192, 193, 200

AWAKE 40, 43, 103, 147

B

backoff 54, 207

    CAP 56, 61, 158, 211, 212

    open MCTA 58

    slot 58, 61

    window 56, 288

base rate 10, 29, 38, 159, 160, 216, 220, 277, 282

beacon 26, 31, 32, 33, 43, 52, 64, 66, 71, 89, 103, 108, 110, 111, 113, 118, 123, 126, 134, 135, 136, 151, 153, 154, 157, 169, 170, 179, 197, 284

    announcement 125, 129, 130, 131

    extension 53, 161

U

UWB 10, 60, 187, 203, 204, 205

W

Wake Beacon

    Interval 109, 110, 115, 122, 125, 126

wake superframe 103, 119, 120, 121, 125

Weaver architecture 255

wide area network (WAN) 23, 261

wireless personal area network (WPAN) 1, 13, 24, 31, 36, 38, 39, 158, 253, 278, 283, 284, 287

Lightning Source UK Ltd.
Milton Keynes UK
UKHW02f1440121217
314351UK00007B/439/P